QUANTITY SURVEYING TECHNIQUES
TECHNIQUES
New Directions

THE EDITOR

Professor Peter Brandon is Head of the Surveying Division at the University of Salford and was previously Head of the Department of Surveying at Portsmouth Polytechnic. He is well known nationally and internationally and has written and presented widely in the fields of building economics and the application of computing to construction. His standard text, written with Dr Douglas Ferry, *Cost Planning of Buildings* is now in its sixth edition.

He is Chairman of the RICS Research Steering Group for all divisions, a member of the SERC Building Sub-Committee and a member of the SERC Information Technology Panel for the Environment Committee. He was Research Director of the DTI/Alvey sponsored research programme for the RICS into the strategic planning of construction projects, which has produced the first major commercial expert system in construction management for the construction industry. He was also Research Director for the RICS sponsored research into Integrated Databases featured in this volume. This book is one of seven of which Professor Brandon is editor or co-author.

QUANTITY SURVEYING TECHNIQUES
New Directions

Editor

Peter S. Brandon

b

**Blackwell
Science**

Blackwell Science Ltd
Editorial Offices:
Osney Mead, Oxford OX2 0EL
25 John Street, London WC1N 2BL
23 Ainslie Place, Edinburgh EH3 6AJ
238 Main Street, Cambridge
    Massachusetts 02142, USA
54 University Street, Carlton
    Victoria 3053, Australia

Other Editorial Offices:
Arnette Blackwell SA
    1, rue de Lille
    75007 Paris
    France

Blackwell Wissenschafts-Verlag GmbH
    Kurfürstendamm 57
    10707 Berlin
    Germany

    Feldgasse 13
    A-1238 Wien
    Austria

First published 1990
Reissued in paperback 1992
Reprinted 1994, 1995

Set by Setrite Typesetters Ltd,
Hong Kong
Printed and bound in Great Britain
at the University Press, Cambridge

DISTRIBUTORS
Marston Book Services Ltd
PO Box 87
Oxford OX2 0DT
(*Orders*: Tel: 01865 791155
          Fax: 01865 791927
          Telex: 837515)

USA
Blackwell Science, Inc.
238 Main Street
Cambridge, MA 02142
(*Orders*: Tel: 800 215-1000
           617 876-7000
      Fax: 617 492-5263)

Canada
Oxford University Press
70 Wynford Drive
Don Mills
Ontario M3C 1J9
(*Orders*: Tel: 416 441-2941)

Australia
Blackwell Science Pty Ltd
54 University Street
Carlton, Victoria 3053
(*Orders*: Tel: 03 347-5552)

A catalogue record for this title
is available from the British Library

ISBN 0–632–03297–9

Library of Congress
Cataloging-in-Publication Data

Quantity surveying techniques: new
directions / editor, Peter S. Brandon.
        p.   cm.
    Includes bibliographical references
    and index.
    ISBN 0–632–03297–9
    1. Building – Estimates.   I. Brandon,
P.S. (Peter S.).
TH435.038   1992
692′.5—dc20   92–6524
                                    CIP

Contents

Contributors' Biographies

Brian Atkin gained his Bachelor's degree in building economics, his Master's degree for research into aspects of the planning of major developments and his Doctorate for research into the time-cost planning of construction projects. He has been awarded a number of major research grants in the field of CAD and applied information technology. Brian Atkin is presently a senior lecturer in the Department of Construction Management, University of Reading where he specialises in technology. (*Chapter 8*)

John Bennett, DSc, FRICS, is Professor of Quantity Surveying at the University of Reading where in 1986 he was awarded his Doctor of Science degree. He is author of many papers including the RICS report 'Construction Management and Chartered Quantity Surveyor' and of the book *Construction Project Management*. He is editor of the *Construction Management and Economics* journal which publishes full reports of QS research. He has travelled to many countries to study international construction practice. (*Chapter 10*)

Geoffrey H. Brown, FRICS, has been a partner of Monk Dunstone Associates, Quantity Surveyors for 20 years and was the Chairman of the Group for the last three years. In 1984 he was invited by the Quantity Surveying Division of the Royal Institution of Chartered Surveyors to chair the Alvey Research Team in collaboration with Professor Peter Brandon, Head of the Department of Civil Engineering of the University of Salford, into the application of Expert Systems to Quantity Surveying Techniques. (*Chapter 7*)

Robert Davidson is a Chartered Quantity Surveyor and received the degree of Master of Science from the University of Reading in 1978. He is currently Head of Building and Quantity Surveying with British Telecom. He has a substantial background in computing and includes Computer Aided Design and Integrated Databases amongst the research projects to which he has contributed. (*Chapter 8*)

Brian C. Edgill, FRICS, is Head of Professional Computing in the Property Services Agency, responsible for the provision of IT systems for the Agency's architects, civil and mechanical and electrical engineers and quantity surveyors. He was chairman of the Steering Committee, set up by the Royal Institution of Chartered Surveyors to oversee the University of Salford study into Integrated Databases for quantity surveyors. (*Chapter 9*)

Roger Flanagan is Professor of Construction Management in the Department of Construction Management, University of Reading. Prior to an academic career he worked for consultant quantity surveyors and general contractors. He has experience of working in the UK, USA, Canada, Mexico, and the Middle East. He has co-authored books on life cycle costing, risk management, the Japanese construction industry, the US construction industry, and Building for Joint Ventures in China. (*Chapter 5*)

Anthony Grice, FRICS, is the Group Technical Development Manager for Bucknall Austin plc. He qualified as a Chartered Quantity Surveyor in 1972 after studying at Leicester Polytechnic and has 25 years' experience within the quantity surveying profession. He was awarded the Earp-Thompson Prize by the RICS in 1971. Since 1987 he has been responsible for formulating technical policy and implementing technical information systems throughout a national network of seventeen offices within Bucknall Austin. Recently he has guided Bucknall Austin through to BS5750 Quality Systems registration. (*Chapter 10*)

David Hoar, FRICS, is Directing Surveyor at Nottinghamshire County Council, with responsibility for Quantity Surveying and Building Economics Services. He was Chairman of the Life Cycle Costing Working Party of the Royal Institution of Chartered Surveyors from 1984 to 1987, which produced two publications on Life Cycle Costing. He is currently a member of the Quantity Surveyors Divisional Council of the Royal Institution of Chartered Surveyors and a member of the Management Committee of the Building Cost Information Service. (*Chapter 6*)

John R. Kelly joined George Wimpey & Co Ltd in 1966 and worked as a quantity surveyor on a number of major building and civil engineering projects. He received his degree in Building Economics and Measurement from the University of Aston in Birmingham in 1970. Prior to his appointment as a Research Fellow at the University of Reading, John Kelly worked for a small architect's practice and the chartered quantity surveyors Davis Belfield & Everest. In 1979 he moved to Heriot-Watt University where he is Senior Lecturer and Course Tutor in Building Economics and Quantity Surveying. He received his MPhil from the University of Reading in 1983. (*Chapter 3*)

John A. Kirkham received a BSc(Tech) in Applied Physics from the University of Wales Institute of Science and Technology in 1966 and the MSc in Business Studies and Operational Research from the University of Warwick in 1972. He has worked in the area of computing for over 25 years both in industry and higher education in the UK and the USA. He is presently a Senior Lecturer in the Information Technology Institute at the University of Salford. His main interests are in Information and Software Engineering Methods, Computer Aided Software Engineering (CASE) Tools and Distributed Databases. (*Chapter 9*)

Philip Marshall received his degree in Estate Management from the University of London in 1960, and is a Fellow of the RICS. He is currently engaged in a research project on behalf of the Rating and Valuation Committee into development valuation techniques, and has so far held interviews with development companies holding assets in excess of £30 Billion — the results of some of his findings will be found as part of *Chapter 2.*

George Norman received the MA degree in Economics from the University of Dundee in 1972 and a PhD degree from the University of Cambridge in 1977. He is currently Tyler Professor of Economics and Head of Department of Economics at the University of Leicester. His research interests cover industrial economics, industrial organisation and investment appraisal. He is the author of a number of books on life cycle costing. (*Chapter 6*)

Brendan Patchell is a Chartered Quantity Surveyor and Regional Associate Director for Bucknall Austin plc in Birmingham. In 1982 he was involved in the evaluation of Bill of Quantities software and became computer manager. In 1985 he wrote a building cost model programme followed in 1988 by a cost planning programme. He is currently working on artificial intelligence and CAD integration in relation to cost estimating. (*Chapter 4*)

Russell Poynter-Brown ARICS is an Associate of Dearle and Henderson, a broadly based practice of Chartered Quantity Surveyors. He is a member of the QS Division's Steering Group on Value Management and represents JO(QS) on the Division's Research and Development Committee. Within Dearle and Henderson, he has particular responsibility for maintaining and developing the Practice's skills in areas such as Life Cycle Costing, Risk Analysis and Development Appraisal as well as heading the Research and Information section. (*Chapter 3*)

Martin Skitmore is a Fellow of the Royal Institution of Chartered Surveyors and an Associate of the Chartered Institute of Building. He has spent 13 years in Quantity Surveying practice in both public and private sectors and 4 years as a systems analyst/computer programmer. Since joining Salford University staff as a lecturer in 1978, Dr Skitmore has been researching a variety of topics associated with the field of construction economics and procurement including his well known work in Quantity Surveying estimating expertise and contract bidding.

He has published many academic papers and reports including a recent book on contract bidding. (*Chapter 4*).

Sue Stevens is a Chartered Quantity Surveyor working in the commercial property sector of the building industry. She has previously worked in private practice and in large commercial organisations in the role of in-house consultant. Before that, she was engaged in three years' full-time research into estimating and cost planning in practice sponsored by the PSA and carried out by the University of Reading. (*Chapter 5*)

Joyce E. Stockley received her BSc Honours degree in quantity surveying from the University of Reading in 1976. In 1986 she joined the RICS/ALVEY Research Team, based at the University of Salford, to investigate the potential use of expert system technology in quantity surveying. She is currently General Manager of IMAGINOR Systems, a partnership formed between the RICS, Salford University Business Services Limited and the University of Salford to develop commercial expert system applications for the construction industry. (*Chapter 7*)

Alan Yates, DipBE ARICS, is a Partner of Dearle and Henderson, a broadly based chartered quantity surveying practice. He is a member of the RICS QS Divisional Council, Director of RICS Journals Limited and a member of the Advisory Panel to the Institution and College Conferences. He is also Chairman of the Construction Management Advisory Committee of the Construction Industry Research and Information Association (CIRIA), principal author of Dearle and Henderson's publication on management contracting and joint author of RICS publication *The Appraisal of Capital Investment in Property*. (*Chapter 2*)

Foreword

The Charter under which the Royal Institution of Chartered Surveyors operates and which gives it both its status and its purpose obliges it, *inter alia*, 'to secure the advancement and facilitate the acquisition of that knowledge which constitutes the profession of a surveyor'. In the practical sense in which this must be interpreted and applied, the acquisition and advancement of knowledge may not be regarded as an end in itself; rather, knowledge refined and tempered by experience must be seen as the fundamental bedrock on which all the skills and techniques of the profession and their development must be based. It is through the application and interaction of relevant and up-to-date knowledge with state-of-the-art skill and technique that a further obligation imposed by the Charter is met, that of maintaining '. . . the usefulness of the profession for the public advantage'.

While knowledge, skill and technique development may be seen as a duty on the profession inherent in the Charter, it also makes common sense for the individual member; it is an imperative to survival in a constantly changing world, a world driven by an ever more demanding market place epitomised by increasing expectations of quality and performance.

It is against this background that the Quantity Surveying Division of the Institution, through its Research and Development Committee, has promoted and invested in many research and development projects over the years. This commitment was consolidated recently through the adoption of a specific policy, the declared purpose of which is 'to promote the scientific, professional and technical R&D activities required to support the changing and developing role of the Chartered Quantity Surveyor in the wider context of the role of management functions and management consultancy as a whole'.

Removed as the profession is, however, from the scientific world of primary knowledge development, its investment must produce tangible results; its R&D programmes must result in new techniques, better skills, wider opportunities. The results of R&D must be capable of being applied in practice. It is this realisation of the application of research results to practice that is the very purpose of this book. Each of the several subjects is dealt with from a dual point of view — the theory behind the respective skills and techniques and their practical application. This is a unique approach and illustrates the benefits of effective and co-operative interaction and collaboration between academia and practice to the benefit of the profession and its individual members.

I commend the editor and the authors for presenting an exciting challenge to all chartered quantity surveyors. I recommend the book to all those in practice, whatever the sphere; to all who are responsible for the education and formation of the next generations; to all who are responsible for the planning and execution of CPD programmes; and finally, to those charged with the task of promoting the image of a dynamic and relevant profession.

Noel H. McDonagh
President, QS Division of the RICS
Former Chairman, R&D Committee of the QS Division

Chapter 1

Quantity Surveying Techniques — New Directions
The Challenge and the Response

PETER S. BRANDON, *Surveying Division, University of Salford*

THE CHANGING ENVIRONMENT

Recent years have seen a rapid growth in research interest within industry, commerce and the professions. Much of this interest has been stimulated by the advances in technology which have forced consultants and manufacturers to reconsider their traditional approaches in order to harness the potential of the computer. The rate of change has produced instability in many industries because no one is quite sure where the next technological breakthrough might appear. An investment in one technology may see it become redundant, even before it is implemented, by a greater advance in another field.

The conservative attitude of the construction industry and its allied professions has prevented major shifts to date but nevertheless the signs are there for rapid change in the future. Christopher Evans (1979) in his popular book *The Mighty Micro* gave a whole chapter to what he perceived to be 'The Decline of The Professions'. He saw the professions as 'exclusive repositories and disseminators of knowledge' with carefully controlled mechanisms (e.g. rules, byelaws, examinations) for preventing access to that knowledge. In some cases the professions refuse to recognise someone who does not conform to membership rules even though they have gained the equivalent degree of knowledge. He argued that, with the development of inexpensive computers and information technology, more of this knowledge would come into the public domain and the mystique surrounding the expertise would begin to disappear. In time this would lead to a demise in the status of such professions.

In the period since *The Mighty Micro* was written there have been significant developments which suggest that the forecast may already be coming partially true. For example, much of the mundane work of quantity surveyors and architects, such as working up bills of quantities and preparing production drawings, is now being shifted to the machine.

Tasks which sometimes came within the professional remit are now clearly labelled as technician work — a step towards the public domain. On the horizon are expert systems which will begin to model and support design and construction decision making. This represents a first low level move towards an 'intelligent' computer which might eventually affect white collared workers in a similar way to how the industrial revolution affected blue collared workers in the past.

THE PROFESSIONAL RESPONSE

Most of the professions have recognised these forces and are moving towards a management role. Management is seen to be the high ground which is less vulnerable to attack from the new technology and in addition provides the control over technology in a competitive world. The next 20 or 30 years will see an interesting power game develop between all the professions for this management role. In construction it is likely to be played out between accountants, architects, quantity surveyors, general practitioners, contractors, engineers and lawyers. It is not difficult to determine the attributes required for a win.

The profession that can offer a 'one step' service for its client, with the widest possible expertise available at its command together with the most advanced techniques and information handling facilities, will prove to be the most attractive. The latter aspect will inevitably mean that this profession will need to be at the forefront of the development of computer applications in the industry. The technology itself will provide a vehicle for uniting the various specialisms found within construction and formal or informal amalgamation of disciplines is likely to follow. It is in this climate that construction research and development is being pursued at the present time.

For many years the Research and Development Committee of the Quantity Surveying Division of the Royal Institution of Chartered Surveyors (RICS QS R & D Committee) has been undertaking a programme of research to provide a better service for quantity surveying clients. By implication it has been moving towards an improvement in its knowledge base for the exercising of a more overt management role.

This book is an attempt to bring to the attention of quantity surveyors and others in the construction industry the benefits of the work that has been done. If the profession is to improve its services and retain its status then research must permeate its culture and be a major part of its role. In addition there must be a willingness on the part of the practitioners and researchers alike to implement, test and experiment with the work which has been undertaken for its benefit. The topics chosen in this volume are those which can either be implemented immediately or which point the way to advances which it is believed will lead to practical implementation in due course. The aim has been to focus on 'how to do it' rather than on the theoretical basis of the work although both are, of course, equally important. Indeed practical implementation without an understanding of the theoretical basis could lead to severe problems in application at a later stage. In particular an understanding of the underlying assumptions, methodology and limitations of a technique are essential if it is not to be applied in an incorrect manner. Each chapter includes references for further reading and it is suggested that these should be followed up to widen the reader's knowledge of the subject area.

THE CONTEXT

The purpose of this introductory chapter is to place the techniques into a context both within the industry and within the policy of the RICS Quantity Surveying Research and Development Committee. In recent years it has been recognised, both theoretically and in practice, that the key decisions are made in the very early stages of the design process even before pen has been put to paper.

The investment in computer technology has reinforced this view with the initial focus being placed on those areas which consumed the most time but were of a more mechanical nature. Such operations as draughting production drawings, sorting and collating NEDO indices or computing critical path networks were some of the first to be tackled. The aim was not only to become more efficient or cost effective but also to release further time for the more fundamental strategic planning and creative aspects of the project. These occur during and immediately after the decision to invest. In the past this is the period of the development/ design process that has had less man-hour input and yet the decisions made have a major effect upon all which follows. The task is therefore to discover techniques, procedures and information support that will improve decision making at this critical period. Most of the topics in this book are focused on this initial part of the development process.

IMPLEMENTATION OF RESEARCH

It has not been the policy of the RICS QS R & D Committee to invest in fundamental research within its subject area but to concentrate on implementation. Fundamental research is usually extremely expensive and time consuming and beyond the resources of a professional institution. However the RICS has an important role to play in disseminating the knowledge obtained through fundamental and applied research and bringing it to the attention of its members and the industry. In part this may involve research into how the work can be used in practice effectively and economically.

Lester Thurow (1980) in his book *The Zero-Sum Society* uses the analogy of road building to identify three different types of research. In road building the first step is to scout the landscape and survey the terrain to see where you want to go and find the best possible route for reaching desirable objectives. This is the role of *scientific research*. Generally scientific research proceeds far ahead of the rest of the road building operation.

It can take several decades before the next stage begins. Well behind the frontiers of scientific research lies the domain of *engineering research*. The direction to take and the basic principles of how to get there are known but a practical road must be designed. When engineering research has been completed, techniques, products and processes move from the domain of the theoretically possible into the domain of those processes which have been mastered and can actually be done. While scientific and engineering research are both important they do not have an impact on professional or industrial practice until their products can be implemented at an economical price. The third type of research is therefore *implementation research*. The landscape may be known, the road can be built, but the road won't be built unless the economic benefits from having the road are greater than the costs of building it. New knowledge only becomes relevant to our economy when costs have been reduced to the point where the information can produce goods and services, which we want, at a price we can afford to pay. Many of the techniques developed in recent years are not used in practice because, *at the present time*, they are too expensive to implement, e.g. the data is too expensive to collect and maintain or the computer power is too expensive or requires too much specialist expertise. It is this task of making research findings practical and economical that has been the role identified for the RICS QS R & D Committee.

INERTIA IN IMPLEMENTATION

A problem with implementation is the inertia created by the existing infrastructure. Japan and West Germany were able to rebuild almost from scratch after the Second World War. Their economies grew from a new industrial base whereas others attempted to adapt from an outdated set of values and equipment. The difficulty of moving this basic set of given assumptions in any economic development inevitably slows the process down.

This may well be the case in professional development and it could be that those countries/professions which do not have the 'hang-ups' of the past may well overtake the current establishment. As world communications grow it could prove to be the service sector of the leading economies, as well as their industrial base, which is threatened by external competition. As stated earlier, even within the United Kingdom at the present time the battle between the professions for the management role in the construction industry is beginning to develop. On an international scale those countries with structures unfettered by the past may well be able to offer a service to clients beyond the scope of the home professions.

All professions survive by the uniqueness of their knowledge and the mastery of the techniques they employ. There is no room for complacency in the foreseeable future. The closer a profession or industry can be to the frontier of knowledge the greater its chance of survival. In the long term it is likely to be more efficient, more productive and provide a better sevice for a lesser fee than its competitors. This is not always possible to appreciate when engaged in practice where the consultant's attention is focused on getting the next job out. However an objective observer can readily see the trends over time and there are professions in the UK construction industry which have become introverted and retreated from roles they have held in the past, handing them over to others. The argument for research and development is a strong one and one that has been endorsed by the RICS QS Division.

RESEARCH DEFINITION

Defining research is a hazardous business and has been the subject of debate in academic circles for many years. It is now interpreted by some institutions in a very broad sense and can include some of the more speculative and creative types of consultancy.

However the following statements may give a flavour of some of the views currently held:

- *A systematic enquiry to test hypotheses* and further the development of both theory and methodology applied both to the procedures involved in the profession and to the substantial issues.
- *The acquisition, dissemination and application* of knowledge, skills and techniques.
- *The collection of data and analysis* in problem solving.
- *The acquisition and analysis of knowledge* in such a manner as to *develop new ideas, methods, skills and techniques.*
- The identification of variables and the *determination of their relationship.*

The key aspects of research are shown in italics and it would be wrong to assume that even this list is exhaustive. Indeed the research brief is extremely wide. It can include reviewing existing knowledge, describing some situation or problem in an effective way for future resolution, constructing something novel or explaining a situation/occurrence more effectively or more robustly.

Professor Grinyer (1981), Head of the Economics Department of the University of St Andrews, classified research according to its nature and type of contribution to knowledge. He suggested four types:

- pure theory
- testing of existing theory
- description of the state of the art
- specific problem solution.

What is interesting about this typology is Grinyer's view that in the field of management PhDs, the opportunity for a truly original contribution to knowledge *decreases*, whereas the chance of a successful completion *increases*, as we go down the list! His discussion relates principally to doctoral studies in management; nevertheless it raises questions that are important in other fields and at other levels of research. In the context of RICS quantity surveying research the last two items on the list are likely to be of most relevance and it is unlikely that earth shattering theory will stem from the profession.

QUANTITY SURVEYING RESEARCH

Quantity surveying as a discipline is difficult to define. It is an amalgam of several other disciplines within a unique context. It embraces economics, law, accountancy, management, mensuration, information technology, construction and so on, all within the framework of the construction industry. Even this framework has been breached with some quantity surveyors exercising their skills outside of construction, e.g. in shipbuilding.

The origins of its techniques are nearly always to be found within these other disciplines but adapted to the particular needs of its construction clientele. Much of its research therefore is focused on applying concepts found elsewhere to the peculiar problem of the construction industry. Occasionally it may find itself in the forefront of the application, as with expert systems, but in other areas the path has been well trodden elsewhere, e.g. value analysis or life cycle costing. Nevertheless quantity surveyors have been in the forefront of implementing such techniques within the construction industry. In addition they have done much to codify and standardise knowledge within many of the procedural systems available today. For example the standard method of measurement, the Building Cost Information Service (BCIS) standard list of elements and the standard form of contract, have had entire or significant quantity surveying input.

Many of the techniques employed, outside of the straight accountancy function, are concerned with prediction, usually of cost or time. It is worth noting the rather different set of circumstances faced by quantity surveyors operating predictive techniques within construction and development. For many reasons the prediction of the future is more difficult than with other industries. Uncertainty is the order of the day. The site, workforce, contractual relationships and so forth vary considerably and the degree of control able to be exercised is also less than could be expected in other manufacturing industries. Consequently not only is prediction difficult but also the management to be able to fulfil a prediction. Indeed even the client can change several times through the development cycle affecting the management strategy.

The complexity is also vast, involving scores of different professions, trades, gangs, suppliers, sub-contractors, supervisors etc. all interacting with one another. There appears to be a parallel here with sub-atomic physics where there seems to be chaos at the sub-atomic level. Matter appears and disappears and moves in a random way yet the object to which the atoms belong is usually consistent, logical and contains order. So in construction, out of the complexity and apparent chaos comes order. If it were not so then prediction would be impossible. Nevertheless the prediction concerns the results of human behaviour and by its very nature this is difficult to forecast. People act differently in different circumstances and it is not possible (or of course desirable) to treat them like machines. There will never be, therefore, complete accuracy and precision in construction forecasting. Risk and uncertainty are the order of the day and this needs to be addressed in the techniques which are employed.

Another problem faced by quantity surveyors is the nature of the information they handle. In the majority of instances it is confidential and belongs to a third party, usually a client. The data on cost and time are therefore not in the public domain, nor ever likely to be so because of their sensitive nature.

When information is collected it is usually held within a single organisation or 'laundered' in such a way that the build-up and detailed content is obscured. With large scale, infrequent products, such as buildings, it is seldom easy to gain substantial samples of detailed information even from a sizeable practice. Without the sharing of such data many of the mathematical models relying on statistical analysis become problematical.

A further problem is the way in which the vast amount of decision making by consultants, contractors, clients, tenants and others interacts so that one person can rarely make a decision without relying upon or affecting another.

The systems which have evolved over the years have attempted to clarify responsibilities and define relationships between parties through the various procurement methods and their respective contracts. Nevertheless it is only necessary to look at the degree of litigation in the industry to see the loopholes and problems. The inter-relationships between consultants, often geographically and sometimes culturally apart, create special problems, many of which are concerned with information. In such a complex process as development, and such a complex product as a building, the update and exchange of information is a major issue. The need to harness the tools of information technology is now becoming paramount.

These issues have been addressed by the research community and can be seen in some of the trends in research topics and the chapters contained within this book.

QUANTITY SURVEYING RESEARCH TRENDS

The foregoing discussion has provided an outline context for the work of the RICS QS R & D Committee.

Over the past decade there have been significant shifts of emphasis in quantity surveying research which can be summarised as follows:

- *A move to the front end of the development process*
  More and more quantity surveyors are taking the lead consultant role, acting as the first point of contact between the design/construction team and the client. This demands a change of technique when forecasting, involving more understanding of how the client's brief can be interpreted and a different kind of assessment of future possibilities using less quantitative information than has previously been provided. It also demands a higher level of management skill if the quantity surveyor is required to continue his involvement into project management.

- *An acceptance of uncertainty and risk*
  In the past most forecasting was of a deterministic nature. For the reasons already given this view has changed and many quantity surveyors are engaged in assessing risk, not only in estimating but in deciding such matters as whether to invest, whether to go to litigation, which alternative yields the greatest profit, etc. The new orthodoxy is to accept risk and uncertainty and to use and develop techniques which take it into account.

- *Taking the longer term view*
  In parallel with the increasing social perception of the need to conserve and avoid wastage of non-renewable resources, there has been a demand to consider total life cycle costs when planning new buildings. Much work has been undertaken in developing a standard classification system and methodology for buildings which should encourage the use of the technique in early development planning.

- *The harnessing of computer based information technology*
  Perhaps the greatest motivator for change has been the development of the computer, together with a massive drop in price which has put it within reach of all. Alongside this development has been a revolution in electronic communication providing the means for machines to exchange information. The basic electronic infrastructure for the free exchange of information is close to completion. The professions, especially those who operate as a team, can ignore this development at their peril. At the moment there appears to be a series of *ad hoc* developments. At some stage a formalised system, into which all can refer and relate, needs to be constructed. It will need to identify the means of integrating information between models, processes, professions and clientele.

- *The reappraisal of manual techniques*
  In parallel with the growth in computing power is a necessary review of our traditional techniques. These have evolved into their present form in order to overcome the deficiencies of human beings. The fact that humans cannot hold vast amounts of structured information with faultless recall, have slow limbs, cannot input or output information quickly, and have poor computational ability has affected the way in which the techniques have developed. Most quantity surveying techniques now need to be reviewed in the light of the removal of these constraints by the introduction of computing power. Much research is being oriented towards this end.

- *The identification of market opportunities*
  The variety of work undertaken by quantity surveyors has increased considerably in recent years. This coupled with fee competition, the growing competition from other professions, the creation of public companies for quantity surveying firms and the general shift towards a more entrepreneurial stance has led to a more market orientated, opportunist approach to quantity surveying activity. The research required to understand that market is only in its infancy, but as with industry there is little doubt that this will be a major influence on research activity, particularly that undertaken in collaboration with firms, in years to come.

These generic trends can be seen in a wide range of research activity across the construction industry. More particularly they can be identified within the research projects in this book representing some of the work of the RICS QS R & D Committee.

INFORMATION TECHNOLOGY AND THE NEED FOR AN INTEGRATED APPROACH

The development process can be considered in four phases, each reflecting a different focus, although each concerned with the project before or after it has been built. Figure 1.1 shows this in diagrammatic form, although the time scales are of course quite different.

In practice there is extensive overlap between the phases depending on such matters as the procurement path, the degree of phased delivery, etc. Each phase has its own particular needs but there is considerable common ground. Continually through the total process the same information is being used, manipulated and processed in some way. For example, a specification level is required at feasibility stage, which needs to be refined during design, built to standard in the construction phase and then accessed for maintenance and other characteristics during occupation.

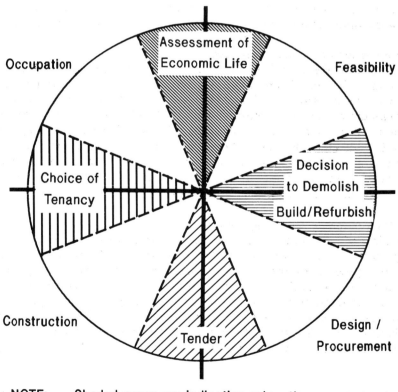

NOTE: Shaded areas are indicative only – they may extend in either direction

Figure 1.1 The development cycle

A similar item of quantitative information, say floor area or functional space, is also used repeatedly throughout the process. The problem is that the same information is held by different people (consultants, clients, contractors, etc.) at different times, in different places, and in a different format. Bearing in mind the sheer quantity of such information, which must run into many hundreds of thousands of items, it is no surprise that failures in communication arise; the same item is recreated or measured within the information systems of several individuals many times; and there is no common source to go to at any time to retrieve information for practice or research. The process is a continuum, yet the information generated is very often treated by consultants like personal property to which others must not gain access. The end result can be a disservice to the client caused by a lack of the right information at the right time.

From a research viewpoint it would seem essential that we address the framework of the total system and endeavour to fit our models and procedures into this framework. If we do not, then much abortive work will be undertaken and the rewards of sharing information will be lost. It will require particular concentration on the project database and the way in which this will link to other databases for such items as technical information, office accounting, cost information, etc.

The technology is available now and we cannot afford to develop our work in a random way assuming manual methods will be employed.

It will be interesting to see in the future whether it is computer aided design (CAD) or management information systems (MIS) that drives the move towards integration. CAD systems are ostensibly large data bases and it may well be possible to attach management routines to them. Alternatively MISs should provide for all aspects of office and project management and encompass the information held within CAD systems. The profession that builds the technology to drive the integration will be in a strong position to retain and enhance its status in years to come.

THE RICS QS DIVISION RESEARCH AND DEVELOPMENT PROGRAMME

The work presented in this volume represents research undertaken through, or in conjunction with, the RICS over the past three years. The reader will note that the topics included are almost exclusively within the first two sections of figure 1.1, i.e. feasibility and design/procurement.

In the context of the preceding discussion the chapters reflect the key techniques employed by quantity surveyors in responding to the new trends as follows:

Chapter 2 — Development and
              Appraisal   — the front end decision to invest
Chapter 3 — Value Management   — optimisation of the benefit/cost
                                   relationship
Chapter 4 — Estimating and   — cost prediction and reliability
              Bidding Methods
Chapter 5 — Risk Analysis   — coping with risk and uncertainty

Chapter 6   —   Life Cycle Costing   —   the longer term view of the cost of the asset

Chapter 7   —   Expert Systems   —   qualitative reasoning to support expert decision making

Chapter 8   —   Computer Aided Design   —   relating the design database to the QS function

Chapter 9   —   Integrated Databases   —   designing the database for efficient and reliable information retrieval

Chapter 10   —   Procurement Systems   —   assessing the contractual relationship/arrangement between parties

Each chapter attempts to encapsulate a description of the technique in terms of 'how to do it' and then provides a short case study. Obviously in the space available it is not possible to do justice to any one of these very large subjects. Nevertheless we trust that the reader will learn something new and have his or her appetite wetted for further study and investigation. All of the authors have produced a more extensive publication list on their subject area and these will provide a good starting point for future work.

The following comments on the techniques are the editor's personal view and not necessarily shared by the authors. They are an attempt to provide the link between the earlier discussion and what follows in the rest of the book.

DEVELOPMENT APPRAISAL

The cost side of the decision to invest has been traditionally the focus of attention of the quantity surveyor. As his management role moves forward to the front end of the development process he must become more aware of the other side, i.e. that of benefits. The benefits relate to the wants and perceived needs of the client, and possibly his tenants, and are reflected in the income derived from the property. Good location and good design should result in a higher income potential. Understanding the relationship between these factors is an important aspect of the entrepreneurial process of development and requires an in-depth knowledge of the property market at both national and local level. That understanding has traditionally come from the general practice surveyor and this is likely to continue for some time to come.

However, as more knowledge about these matters moves into the public domain, so it will be possible for quantity surveyors and others to make that judgement. The techniques in the chapter by Yates and Marshall are well established and need to be mastered by a quantity surveyor if they are to perform the appraiser's role or that of the client's representative looking after and monitoring his interests. However, they do require considerable market knowledge which usually comes from experience. It is not therefore merely a case of fitting standard figures into the variables in an equation. Knowledge of the needs of a locality, land use in the area and all the other factors affecting supply and demand are essential for quantifying the income side of a revenue producing building. Nevertheless if quantity surveyors are to provide the one stop service which many clients will demand for their financial management, this is an area which deserves more attention.

VALUE MANAGEMENT

After development appraisal the activity which follows is that of design from conceptual thinking, through formulation, to a concrete proposal. Once again the quantity surveyor has traditionally followed this process from the cost side of the equation. Indeed in the not too distant past his reputation was built on his ability to cut costs and this subsequently led to a negative image relating to design. Fortunately in recent years he has played a more positive role in suggesting alternatives from which the design team, including the client, can choose. What the client demands is maximum benefit from minimum cost. Inevitably this leads to compromise on both sides and the result from the client's point of view is 'best value' — i.e. the best balance between the resources (cost, time, finance, etc.) and the benefit (income, welfare, function, etc.).

This search for value is a role ideally suited to quantity surveying and provides a much better image for marketing his services than the 'cost cutter' of the past. It is not a role that he can undertake alone. It is a multi-disciplinary function involving all aspects of the design process. However, the management of this value process in a positive manner is suited to the quantity surveyor in a similar way to the building economist and cost engineer in the USA and elsewhere. In these other countries, where quantity surveyors are not found, the emphasis tends to be on the monetary savings that can accrue from value management. In the UK we have the opportunity, because the cost function is well catered for, to gain and re-emphasise the benefit side of the evaluation. The methodology employed in value management, which is really just a logical structured framework for the thought process, is of assistance in making a decision. The techniques of evaluation are already in the quantity surveyor's possession and need only to be enhanced by the assessment of benefits in a more rigorous manner.

Of course there is no guarantee that value management will achieve its objective and some would argue that it could militate against 'creative leaps' and lateral thinking. Nevertheless the approach, by focusing on value, is more likely to make a positive contribution to satisfying clients' requirements than if it were ignored. For quantity surveying it offers a positive image under which it can market its services.

ESTIMATING AND BIDDING

The prediction of the price to be paid by the client is probably the most fundamental professional service provided by quantity surveyors. For generations it has been the backbone of all the profession's activity, whether it be estimates at the feasibility stage, pricing alternatives, bills of quantities or cost monitoring during design and construction. Practically all the models used for this process have been built around the measurement of items 'as placed' in the building and more particularly as classified by the UK Standard Method of Measurement.

This base information has been reclassified in a number of different ways including area, space, elements, volume, etc. as well as simplified measurement in the form of approximate quantities. In addition it has provided the source data for reclassification into operational activity and the creation of statistical models. There have been very few attempts to provide a complete overview of such models and one of the interesting aspects of Skitmore and Patchell's chapter is the manner in which it attempts to provide a taxonomy for estimating models. It looks at three more recent approaches, i.e. regression, simulation and expert systems, in detail but it is probably too early to judge whether any or all of these three are likely to produce lasting solutions to the forecasting problem.

From the work undertaken at the University of Salford into expert systems it would seem that practitioners wish to blend their expertise with the model if they are to gain confidence in its use. A 'black box' approach, where figures are fed in at one end, results appear at the other, and the consultant is excluded from what goes on in between, does not gain the confidence of the users. The complexity of the process is difficult to capture in a simplified model and yet, if there is no interaction, the model must be assumed to be perfect. Perhaps researchers should address more fully which aspects of the estimating process can best be handled by machine or by the human.

The reliability of estimating models is a problem in its own right. Where building stereotypes are well established and the new building conforms to the existing pattern, then reliability can be high. Where the type, design or location of the building do not conform, then it is much more difficult to assess a future price from past data. These issues are investigated in the paper, although as acknowledged much more work will be required before a robust taxonomy and evaluation of estimating systems can be established. Nevertheless the authors provide an excellent datum from which to start.

RISK ANALYSIS

This subject is probably attracting more attention at the present time than any other, with the possible exception of expert systems. For too long the industry has assumed a deterministic approach to forecasting. 'The client requires a single figure' is a statement often heard in quantity surveying circles and it is probably true. Finance houses, public authority committees and businesses need to know what their commitment is from the very start. They do no want to know that it can fall within a range, despite the fact that most people recognise that you cannot anticipate accurately the final cost of anything as complex as a building over a period of several months and even years. It seems that clients expect their management and design teams to control and monitor costs to the point that they will hit their target. This may or may not be in the client's interest because one sure way of hitting target is to make the target larger in the first place!

Nevertheless a retrospective view of first estimate with final account would show a wide discrepancy on most contracts — even though consultants can explain why! There is a case therefore for educating clients to expect a figure within a range and to inform them of where the risks lie in making the estimate. In other fields too, such as litigation, development appraisal and time forecasts, there is a case for identifying and quantifying the risk. If the reality is uncertainty then it seems strange not to recognise this fact in the information that consultants supply.

This is not to say that management of the risk should be ignored. Once it has been defined and quantified, everything should be done to minimise the risk and to keep the client informed. It does, however, require a different approach to conventional estimating and control procedures. Almost by definition the focus of attention must be on defining where problems can occur and providing a framework to avoid such problems. The chapter by Flanagan and Stevens addresses the analytical aspects and provides a good introduction to one risk analysis technique, that of Monte Carlo simulation. It should not be forgotten, however, that this is merely the start of the risk management process.

LIFE CYCLE COSTING

For some time it has become apparent that those clients, particularly in government, who maintain responsibility for long life built assets require a long term view of their investment problem. Over the past three decades a concern with life cycle costs in one form or other has been evident. Whether it has been called 'cost-in-use', 'terotechnology' or 'total building cost appraisal', the aim has been to look at the total client investment over its lifetime or at least a substantial number of years.

The major problem has been how to forecast so far into the future. To attempt to do so could border on witchcraft because the underlying assumptions are based on enormous generalisations on which it is impossible to rely. A major problem here is the lack of management systems to monitor, control and check against the plan of expenditure. Even if these were in place, it would be impossible to assess all the external factors influencing life cycle costs over such a long period of time. Changes in taxation, differential inflation, technological change, changes in building use, political instability and so forth are not easy to forecast. Past data is of little help, except perhaps on the maintenance of traditional materials, as it is largely 'budget led'. What is done is what can be afforded, not what is needed.

The reliance on the discount rate to reflect investment decisions over time is also suspect and it has to be decided whether the technique is an attempt to reflect the real decision making attitudes of the client (if so, it needs much more refinement in terms of risk seeking or aversion) or whether it is a means of social engineering on behalf of governments. Some governments, for example, specify a lower discount rate for energy costs in order to make future expenditure on energy appear as an even greater proportion of total costs. These matters are significant and require much more discussion than is possible in a book of this size.

The authors of the chapter on life cycle costing, Norman and Hoar, have, as requested, focused on how to undertake the evaluation in a form now becoming standardised through the RICS work on the subject. It is important that we address the future more overtly than in the past. The 'green' lobby and environmental concern groups have made us all aware of the rapidly dwindling stocks of non-renewable resources. These issues alongside the question of future costs will no doubt mean that more attention will need to be given to the subject. Some interesting work is now being developed abroad, in the USA, Finland, Netherlands and France, for example. We need to be aware of these developments as we use the technique and we need to begin to develop the management systems for control which may make the life cycle predictions more meaningful.

EXPERT SYSTEMS

A feature of building economics research since the last world war has been the desire to improve on the models which consultants use to make their judgements. The search for improvement has tended to follow mathematical or statistical routes. To some extent these have been successful in providing useful models within narrow domains (e.g. regression) or assisting in the understanding of the problem under consideration (e.g. simulation). The uptake among practice, however, has been relatively small. The reasons are not hard to see and some have already been mentioned in this editorial under the estimating heading. There is a limit to what can be expressed in a straightforward equation and there is an 'understanding' beyond the basic cost variables, which is known to the consultant and which allows him to modify his basic model. The mathematical model is merely a 'skeleton' reference point on which he builds the 'flesh' of his expertise. If part, at least, of this expertise could be grafted on to the base model it would allow less experienced staff to reach the higher level quicker, would provide a check list for the expert and would release the expert to give more time to those parts of decision making in which he is most able. Expert or knowledge based systems are a first step towards capturing the routine knowledge of the consultant and making it available to a wider audience.

The RICS/Alvey research project was a government and profession sponsored piece of work on behalf of the UK construction industry to demonstrate the potential of such systems. It has succeeded beyond what was envisaged by the sponsors and has resulted in the only major commercial expert system for project management in the UK. In this domain it appears to lead the world and papers have been given in more than a dozen countries to date. Indeed it was the only one of the DTI Alvey 'community club' projects to produce a commercial system within two years of completion. This is an area therefore where quantity surveyors have taken a lead in the application of this new technology. With 120 systems sold in the first year it looks as if expert systems are here to stay.

The potential appears to be enormous and as the software improves and knowledge engineering methodologies develop so it is likely that these systems will be easier to produce and will become more widely available. Their transparency in use and their similarity to the decision making of the consultant make them readily acceptable to practitioners as well as a useful educational tool for students. There are of course limitations, and it is still necessary to define what aspects of advice should be left to the machine and what should be left to the consultant. Nevertheless the response from the profession has been good and further developments are in hand to extend the range of applications.

COMPUTER AIDED DESIGN

If there is a technology which strikes at the very root of traditional quantity surveying activity then it is the development of computer aided design (CAD). Since its inception quantity surveying has depended for the majority of its fee income on the measurement for, and preparation of, the bill of quantities (BQ). In essence this is the assimilation of quantities and associated specification of the building, taken from the architect's design drawings and classified in a standard format for pricing by the contractor.

CAD systems can be considered as large databases which hold the same kind of raw information and to which analytical and scheduling routines can be attached. Even with the current state of technology it is possible to schedule large sections of current BQ information in a form suitable for tendering purposes. Currently these schedules will not necessarily conform to the standard method of measurement but nevertheless would provide a fair basis for contractors to tender. Even though there are problems in moving to the measurement of a complete building (including the fact that the designer will need to have his work complete in every detail), most problems could be overcome through the current knowledge of the technology if the financial investment was made.

This strong possibility could revolutionise the structure of quantity surveying firms, tendering arrangements within the industry and status of the quantity surveyor, should he try to hold to his traditional role. Inevitably it will be seen by some to be a threat to quantity surveying activity. It would be wrong to minimise the difficulties, both technical and cultural, in producing tender information direct from the CAD database, but it appears that the will and intent is there. To ignore such an important development would be to miss out on the opportunities it offers. However, it will require a rethink of the role of the profession and the professional structure required to support that role. If quantity surveyors can move in the direction of financial and project management and create the support systems required, then CAD becomes a helpful adjunct to this role. Much design information can be used in management, and the design process itself requires managing. Some quantity surveyors are introducing CAD systems into their office for early feasibility studies and other functions and this is likely to be a growing trend although it would be wrong for the quantity surveyor to usurp the design function. Design is a highly skilled activity which requires a different sort of education. It is likely that quantity surveying practices will buy in the design expertise when they require it and use it alongside their management function. As the edges become blurred between the professions, it could be that many of these techniques will be shared and only a change in emphasis will identify the root discipline.

INTEGRATED DATABASES

Irrespective of the decisions taken by a single profession there is little doubt that the concept of the information society is here to stay and will drive all organisations and bodies to share their data and knowledge. There will be resistance and it will take scores of years to implement. It will not happen in one large operation but piece by piece, industry by industry. However, the technological infrastructure, the 'highways' for the information, are already being built, with Japan leading the way.

Each industry will need to design its own information system, built around a database, to handle its own affairs. Construction is no exception and, with its multiplicity of trades, professions and manufacturers, will be a very complex activity to model. This complexity may actually encourage the development of such systems in order to improve the efficiency of the industry. In addition the system will need to link to the wider world of industry and commerce of which it is a subset. As can be imagined, it would be impossible to tackle such a task in one go.

In the first instance an agreed framework is required within which to work. The RICS has taken a commendable initiative in beginning to specify this framework which should lead to major advances in the future. This should place the profession in an influential position to suggest the form and content of the system and play an important role in its development in the years that lie ahead.

PROCUREMENT SYSTEMS

It would be true to say that quantity surveyors have been at the centre of the development of new procurement systems. Their knowledge of building contracts, project management and contract documentation has given them clear insight into the possibilities of how the arrangement of these contractual relationships between parties can benefit or hinder the client. Until recently the judgement on which arrangement was best was made on an initiative basis from past experience. This experience served the consultant well, although there is some evidence to suggest that procurement methods go through periods of 'fashion' which need rigorous reassessment to ensure that the market does not exploit the inbuilt bias caused by a particular approach being in 'vogue'. There is a sound argument for the choice of procurement being made largely on the grounds of the experience of the consultant managing the process. If he has wide knowledge of one method and not others, then it would seem inadvisable to choose an alternative approach in which he may be forced down a difficult learning curve. The skill is personal and this may be an overriding factor in the procurement choice.

Nevertheless aids to decision making are useful and the paper in Chapter 10 outlines the key factors and suggests a quantitative approach to making a selection. In addition the 'Elsie' expert system mentioned in Chapter 7 includes a module on procurement containing a consensus view of a dozen quantity surveying practices. Both approaches rely on a definition of the key characteristic of the project and a mechanism for weighting these in relation to one another.

The problem with research in this area is the difficulty of validation. It is impossible to build the same building under exactly the same conditions several times using different procurement methods. Some objective testing can be undertaken in respect of the reaction of various parties to the performance of a specified method or a particular project. This does not mean, however, that the same success or failure will be repeated on the next project. The subject remains open for debate and experiment. What the quantitative and qualitative approaches provide is a datum based on the distilled experience of expert consultants. This can then be tested and modified in due course as more information and experience is gained. The personal knowledge of the managing consultant should, however, never be understated.

CONCLUSIONS

It can be seen from the above description of the chapters that a very comprehensive range of techniques is included in this volume. In no way can the contents provide an exhaustive study of each subject or its application. Nor can the techniques themselves provide a panacea for the multiplicity of new problems arising from changing circumstances in the commercial world. Progress is largely by evolution and refinement. The subject matter of this book is an attempt to raise the level of awareness of the quantity surveying profession to the tools that are available for an improvement in its service to its clients. In addition the book seeks to provide the ground rules distilled from the RICS research for applying these techniques in practice. Some of the papers are closer to existing practice than others, but taken overall they provide a guide to those tools which are likely to be used by quantity surveyors in the immediate and short term future.

It is no coincidence that nearly all of the techniques require computer support to be effective. A key priority must be, therefore, education of all members of the profession to ensure computer competence and literacy. As the information revolution gathers force, so it will be those who can operate the supporting technology who will be the most competitive.

It is the integration of information systems, together with their associated techniques, which will offer the greatest opportunities and greatest threat to any of the service professions. These professions rely on their knowledge. With greater access to this knowledge through information systems, a profession can either expand its own knowledge base or attempt to place further barriers to access to its own specialisation by forms of restrictive practice. It is likely that both will be attempted by most professions and it will be a case of where the emphasis lies.

There are a number of strategies that could be employed by a professional institution in this respect, including:

- Fight to defend the knowledge of the profession and seek to raise further barriers to others wishing to undertake quantity surveying work — the philosophy behind a siege economy.
- Seek to enhance the knowledge of the profession so that it can move quickly into new areas as opportunities arise — at the same time be prepared to move away from old methods when technology and competition make them redundant.
- Do nothing corporately but allow individual firms to seek their own place within the market.

Of these, the siege mentality may work in the short term but the barriers will eventually be circumvented. This psychology may also result in a stagnation within the discipline which leads to a lack of entrepreneurial skill and competitiveness.

The expansion of the knowledge base sounds fine in theory but difficult to implement in the short term in practice. In the transitional phase from old to new knowledge the profession would have to accept a drift away from traditional skills to incorporate the new into its educational system. It is impossible to encompass all possible knowledge into an undergraduate scheme and it is difficult to predict what the new knowledge will be in any case. The extent of the new knowledge is likely to be well beyond the current continuing professional development re-quirements of the RICS and may require retraining. To undertake an expansionist role may require a greater degree of leadership by the professional institution than it has been used to in the past.

To forego a corporate policy led by the institution, allowing each firm to develop in its own way, has many attractive sides to it. Motivation will be higher, the incentive greater and the pace of development dictated by the market. However, there are problems, and in particular the size structure of surveying firms suggests that firms may find it hard to provide the investment capital to compete in the new markets. Unless they act corporately in investing in research, they will be in danger of being left behind.

These are large issues and to some practitioners remote from their day-to-day activities. Nevertheless, they will have long term implications for the quantity surveying profession. This book is an attempt to improve the knowledge and skill base required to allow the quantity surveying profession to take the opportunities as they arrive and to respond to new client demands. It will have succeeded if it acts on a catalyst for further study and implementation of the techniques it describes.

BIBLIOGRAPHY

Evans, C. (1979) *The Mighty Micro*. Sevenoaks: Coronet Books.

Grinyer, P. H. (1981) 5th National Conference on Doctoral Research in Management and Industrial Relations. University of Aston and quoted in Howard, K. and Sharp, J. A. (1983) *The Management of a Student Research Project*, p. 12. Aldershot: Gower.

Thurow, L. (1980) *The Zero-Sum Society*, p. 78. Basic Books Inc., European Book Service.

Chapter 2

Development Appraisal

PHILIP J. L. MARSHALL, *South Bank Polytechnic, London* and
ALAN YATES, *Dearle and Henderson, London*

INTRODUCTION

A paper on development appraisal may be considered an unusual companion to a series on quantity surveying techniques and practice. However one reason for its inclusion is that development appraisal is not a technique as such but more a process whereby several techniques are applied and various inputs are involved.

One key input to development appraisal is provided by the quantity surveyor and involves a range of expertise including construction costs, specifications (quality), market trends, programming and procurement. In other words the very techniques and practices covered by the other chapters in this book.

Given the demands now placed on the property professions for vigorous development appraisal and evaluation techniques together with advances recently made it is felt that a chapter reviewing the subject of development appraisal and highlighting implications on the role of the quantity surveyor is appropriate.

Therefore in this chapter, background information on development appraisal is provided followed by a review of the types of appraisal commonly used in the UK development market. A further section contains worked examples of the various development appraisal techniques followed by notes for quantity surveyors on their input for particular aspects of the process.

The chapter has been written for quantity surveyors in consultancy situations to assist them in their role when advising development teams. Quantity surveyors in project management or development organisations with responsibility for appraisals, whilst hopefully finding this chapter of interest, should not regard it as providing the necessary depth for the subject as a whole.

BACKGROUND TO DEVELOPMENT APPRAISAL

Recent publications define the current new era in development appraisal as being born out of the turbulent years of the 1974 property crash and the pre-1980s recession. The period has seen an effective end to any reliance on the short-cut or 'back-of-an-envelope' methods of assessing the effectiveness of development projects (see Darlow 1988).

Prior to this time and especially the years immediately after the Second World War demand for all types of accommodation was high, costs and incomes were reasonably stable or were within reasonable ranges of prediction and virtually any site could be considered for development. Today and for the foreseeable future all these aspects are volatile and subtlety must be employed in appraising their potential.

The developer and other investing parties to a development can now employ a sophisticated array of techniques and data. Also, as property continues to become an increasingly significant part of investment choice (whether or not such vehicles as unitisation/securitisation advance) so the need will exist for appraisal techniques to be in line with those used for other forms of investment.

Further, a recent research project on behalf of the Rating and Valuation Committee of the Royal Institution of Chartered Surveyors highlighted the following important changes in the market, namely:

(a) The growth of software to aid development valuations and appraisals.
(b) The growth of firms specialising in producing tailor-made systems for development clients.
(c) The increase in computer 'literate' surveyors entering into the profession able to use financial spreadsheets.
(d) The intense competition between developers to buy important sites for development leading to the need for a higher level of sophistication in preparing calculations in order to win sites in tender, at auction or by private treaty.
(e) The substantial increase in bank lending to developers (debt finance) especially via syndicated loans arranged over a longer than usual term and also on more favourable terms than previously.
(f) The development of the forward sale transaction in 'profit related' buy-outs of the developer's interest.
(g) The growth of corporate funding techniques such as the issue of deep discounted bonds, sterling commercial paper, 'swaps' and so forth.

Current development appraisal techniques attempt to cater for the above factors. However, in a chapter of this nature it is only possible to discuss general principles. Variations will be encountered in practice due to the requirements of a particular investor or the design of the appraisal software employed.

TYPES OF DEVELOPMENT APPRAISAL

All development appraisal methods have a common purpose; that being to enable a range of values to be calculated by comparing costs with incomes. The skill needed in this process is fourfold namely:

(a) Selecting the method of development appraisal that will be the most suitable for the scheme.
(b) Identifying all costs and income aspects.
(c) Estimating and predicting (b) as realistically as the information will allow.
(d) Presenting the results in a manner that is meaningful to decision makers.

All four aspects are of course inter-related but it is essential that at the outset an appropriate appraisal method is used for the particular stage a development has reached. Very often a combination of methods is used during the life of a development as better information is obtained and especially when a scheme passes from the strategic to the detailed stage.

The main development appraisal methods are commonly classified as follows:

(a) The residual valuation method
(b) The feasibility or residual profit method
(c) The cash flow
(d) Discounted cash flow
(e) Sensitivity analysis
(f) Ground rental appraisals
(g) Forward funding appraisals
(h) Long term funding appraisals

These are now discussed briefly and a summary of their principal advantages and disadvantages is provided in Table 2.2 at the end of this section.

(a) The residual valuation method

Essentially this is the initial assessment of the price to pay for a site (or sites). Thus for a scheme to be viable the value produced must exceed all costs (plus profit). It can be expressed as follows:

Gross development value   [Net sale price(s) or
(before sales costs)   net investment value]

Less all costs of development *plus* interest
add return for profit and risk (not a cost)

Amount (if any) left to purchase site(s)

Less (a) Costs of land purchase ⎫
      (b) Interest on land costs ⎬

= Residual land value

(b) The feasibility or residual profit method (or stack-up)

This can be expressed as follows:

Gross development   Net sale
value   prices or
(before sale costs)   net investment
              value

> *Less* (a) Land costs
> (inc. possession)
>
>   (b) Building costs
>
>   (c) Interest costs on
> (a) + (b)
>
>   (d) Selling costs
> etc.
>
> =   Residual profit

The latter method is probably more common than the straight residual calculation as many sites are being offered to developers at prices which perhaps do not reflect the true 'residual land value' but are instead an owner's optimistic view of the site's worth. Additionally most computer programs are based upon the feasibility method and make use of the same inputs to enable a range of outcomes to be calculated including residual land values if required.

Once this calculation is completed, however, probably the most important return to any developer is the rental yield, since the scheme will either be retained for investment (and so this yield needs to be compared with other yields) or sold on to a fund or other investing institution (in which case it will be necessary to know the net return).

The rental yield is in effect:

$$\frac{\text{Net rentals received}}{\text{Total development costs}} \times 100 = \ldots\%$$

and will normally exceed the equivalent investment yield by at least 1%. For example if a prime office rental yield is 5.75% then the likely development yield will be above 7%. (Note: during periods of intense competition for sites prices for land may be increased to levels at which the net yield falls below these figures.)

(c) The cash flow

This method is basically a more accurate approach to the feasibility method. The profit of a scheme is produced by projecting the cash flows of costs and incomes. The feasibility method itself suffers from the fact that costs are only averaged out over the estimated development period rather than scheduled at the time when they are likely to be incurred. The cash flow method has the advantage of enabling the monthly or quarterly interest charges to be calculated and identifying when they are likely to be incurred. This is an essential requirement of many funders when advancing bridging finance on schemes.

(d) Discounted cash flow

Essentially similar to the cash flow approach in that all likely cash inflows and outflows are projected over the development period. However, instead of calculating interest charges on a month by month 'rolled up' basis the actual inflow or outflow is discounted at the selected target rate (usually the cost of finance or the opportunity cost rate). The resultant 'Net Present Value' (NPV) can either be positive — in which case a profit has resulted — or negative — in which case a loss will be incurred. In the unlikely event of the NPV equalling zero neither a gain nor loss is incurred.

The method can then be extended to calculate the internal rate of return (IRR) on the scheme. This rate is essentially the discount rate (present value of £1) which needs to be applied to the projected series of cash flows so that the total present values (PVs) of the inflows equate with the total of the PVs of the outflows, i.e. the NPV = 0 and neither a gain nor a loss is incurred. The IRR has a number of advantages to both developers and funders:

(i) It enables a comparison to be made between the rate of financing a project and the overall return. Thus if a project is financed at say, 12%, the IRR

Table 2.1 Common matrices

(i)

Yields	Rents		
	£20	£25	£30
6% 6.5% 7%	decreases	Profit as return on costs	increases

(ii)

Building costs	Building time		
	12 months	15 months	18 months
£100 m2 £125 m2 £150 m2	increases	Profit as return on costs	decreases

(iii)

Interest rates	Development time		
	20 months	24 months	28 months
13% 14% 15%	increases	Land value	decreases

should be at least double (before tax) — the difference in yield representing the 'profit yield.'

(ii) The IRR can be used to rank schemes — the higher IRR schemes being given the highest priority.

(iii) The IRR is also used by funds when looking at schemes to finance. Comparison will be made with other investments available at the time having regard to initial yield or projected income growth potential.

(e) Sensitivity analysis

Sensitivity analysis involves the evaluation of the critical variables within an appraisal and realistically should be undertaken using an appropriate program.

For example, a range of rents can be compared to yields, building costs, interest charges or land values. Some of the more common matrices are shown in Table 2.1.

(f) Ground rental appraisals

In many urban development schemes the land owner (or in reality the local authority) will wish to control the timing and quality of a particular scheme as well as share in the long term profitability of the scheme. The usual method is to advertise the scheme and invite developers to tender 'offers' for the scheme.

The basic terms for tendering are usually contained within the 'development brief' and parties will be invited to submit their detailed schemes (with financial offers) involving a combination of the following:

(i) The provision of unproductive works, basically works involving non-income producing facilities such as road improvements, community halls, car parking or public housing.

(ii) A premium payment, usually a part payment on signing of the building agreement and final payments on practical completion.

(iii) A ground rental offer made up of a number of different stages. These stages might be as follows:

Stage (a)   The development period — usually a peppercorn rental liability.

Stage (b)   On practical completion/letting. The payment here will be based on the initial appraisal which is normally carried out early in the development period. A typical list of input variables would be:

Estimated rental income   =   £
Estimated total development
costs
(inc. interest, fees plus

premium offer, if any, etc.).

X return required (% on cost)		%
(a) Yield to fund	=	%
(b) Yield to developer	=	%
(c) Annual sinking fund (to replace parts of buildings)	=	£
Mininum ground rental offer	=	£

Stage (c) On completion. The figures used in (b) will be based on estimates only of costs and rents. The actual figures for a large town centre scheme are likely to be significantly different from such estimates and so although the minimum ground rental will be payable to the local authority an additional share of any 'overage' (improvement) will be payable on a pre-arranged formula basis.

Stage (d) Upon the review of the ground rental in perhaps 3–5 years after completion. The ground rental will be recalculated as a percentage of net rents receivable — again usually worked out on a formula basis.

(g) Forward funding appraisals

Currently many (especially smaller) development companies are entering into forward funded 'profit related' buy-out arrangements. Under such arrangements the developer will receive:

(i) Often 100% of development costs provided by the fund.
(ii) Probably an advantageous rate for interim finance. Usually at rates of between 8% and 9% pa.
(iii) The removal of the main risks involved in the scheme, now borne by the fund (except for any developer's guarantees given).
(iv) At the very minimum a fee for project management even if no profit has arisen.
(v) The certainty of a pre-agreed buy-out price based on final costs and rentals received, even if the yield used in the calculation gives a lower price than actually the property might have achieved if sold as an investment.

The fund in return receives:

(i) A priority yield on the scheme — usually marginally above the comparable investment yield rate.
(ii) A pre-agreed share of overage arising and in certain cases the retention of the whole of the overage income if a limit ('capping') on the developer's return has been arranged.

(h) Appraisal of long term funding arrangements

In certain cases the developer may wish to retain the scheme as a long term investment in which case he may decide to repay short term loans by one of the following methods:

   (i) By sale of the completed and let scheme to an institution and lease back to himself at a head rent based on the fund's required yield. Leases in these situations can usually be between 75 and 125 years and subject to upwards only rent review clauses at either 3, 4 or 5 yearly intervals. Yields would normally fall between 6% and 8% for town centre retail or office schemes although industrial schemes may be funded at slightly higher yields.
  (ii) By way of a long term mortgage secured upon the value of the asset — the interest may be fixed (higher) or variable together with many combinations of methods of repaying the capital.
 (iii) By way of corporate funding which might include the issue of new convertible loan stock, a deep discounted bond issue or similar arrangement. The type and amount of such capital issued will depend upon the stock market status of the development company.

Under such arrangements the initial profit allowed for in the early appraisal is 'taken out' of the scheme and retained as an income profit. When comparing alternative funding arrangements of this type the only realistic method of appraisal is discounted cash flow (DCF). Such an appraisal would be approached as follows:

- Projections would be made over, say, a 50 year term of the rental income arising from the scheme after allowing for rental growth at each review.
- Next a deduction would be made from such income of the interest of the head rent payable to give the 'true return' to the developer.
- Finally in order to find the present value of such projected net incomes each tranche of developer equity income would be discounted by an appropriate discount rate (the developer's target rate).

Over 50 methods of this type of funding are now used in practice and a study of these would constitute a whole book.

A detailed discussion of the case of DCF analysis when comparing alternative funding arrangements is contained in Chapter 9 of Darlow's *Valuation and Development Appraisal* (Darlow 1988).

The principal advantages and disadvantages of the appraisal methods described above are summarised in Table 2.2.

Table 2.2 Selecting the appraisal method — a summary of their principal advantages and disadvantages

Method	Advantages	Disadvantages
(a) Residual valuation	1 Quick, straightforward method of calculating land values. 2 Calculations can be worked manually using only a calculator or a programmable calculator.	1 Small variations in the critical variables can give wide changes in the underlying land value. 2 The method assumes how 'the market' will value the land — in practice bidders will use different criteria in their own appraisals.
(b) Feasibility or residual profit method	1 As above, this is a quick method of 'stacking-up' costs and assessing the profitability of a scheme when the land price is known. 2 When used in conjunction with sensitivity analysis the impact that variable bids for land have on profit can be evaluated.	Same as above
(c) Cashflow	1 Although not strictly an advantage this method is essentially a requirement of any bank financing a project. 2 Identifies when maximum drawings for the project will occur and interest payments are due thereon. 3 Can provide an important aid to cost control	1 Generally a computer is needed to complete a cashflow for a large or complex project.
(d) Discounted cash flow	1 A useful method to help with decision taking. 2 Where the internal rate of return (IRR) is calculated, the method can provide a useful tool for the ranking of projects. 3 A quicker method than cashflow analysis as interest is not added into the cash flow since it is the PV of £1 figure that is used to discount the projected inflows and outflows. 4 Very useful method of assessing alternative tax advantages of a particular project (e.g. receipts of capital allowances in an enterprise zone).	1 Does not show interest payable on project. 2 When information is limited distortions in the IRR can occur.

Table 2.2 contd.

Method	Advantages	Disadvantages
(e) Sensitivity analysis	1 An essential exercise when evaluating the effects of uncertainty (risk). 2 Matrices of rents, yields, interest charges, building costs and profit provide useful tools in achieving the objective of maximising the returns on a scheme.	1 None, but impractical without the aid of a computer for complex variables.
(f) Ground rental appraisals	1 Probably the only way of evaluating a rental payable under an urban development partnership project. 2 Projections can be made of the interaction of the ground rental with rack rental over the life of the project enabling the developer's net income to be evaluated.	1 Ignores in certain cases future obsolescence of the buildings. 2 Procedures for accommodating revisions of ground rental in the event of changing costs and income can give rise to complicated building and finance agreements.
(g) Forward funding appraisals	1 Basically an extension of the feasibility method but alternative buyout values can be appraised and decisions taken accordingly.	n/a.
(h) Long term funding appraisals	1 Enables the developer to assess the benefit of retaining a project for long term investment. 2 Enables a comparison to be made of the alternative funding methods available at any given time and from a wide range of sources. 3 Allows the funding institution likewise to assess its long term 'equated yield' for investing in the scheme.	1 Long term projections of future rental growth difficult. 2 Variations in discount rates used can produce different NPVs (e.g. a higher discount rate will make future returns less attractive than they might really be).

WORKED EXAMPLES

In this section an example is provided for each development appraisal method discussed previously.

The main calculations involved in each appraisal method are summarised in Tables 2.3–2.9. Explanations of these calculations and their underlying assumptions are provided in the notes for each table.

For ease of illustration assumptions and principal data have been simplified. It should be appreciated by the quantity surveyor that in practice a range of issues not covered in these examples will need to be considered by the development team. These will include:

(a) Density limitations and other permitted uses for the site which may also need to be appraised.
(b) Legal constraints, e.g. rights of light, restrictive covenants, rights of support and so forth.
(c) Site development problems; examples from recent projects have involved toxic waste in the subsoil and the discovery of an Elizabethan theatre!
(d) Calculation of net lettable areas from gross building areas, i.e. the 'efficiency ratio'.
(e) Estimation of future rental levels and the impact of the scheme on future rental levels in the locality together with the letting problems which could be encountered.
(f) Methods of funding the scheme and likely movements of interest rates over the development period.

Appraisal task and key data	**Solution**

Task A

Calculate the residual value available for the purchase of a site which can be developed for offices based on the facts below:

Use residual valuation method

(i)   gross building area 100 000 sq ft measured internally;

Refer to Table 2.3

(ii)   the net lettable area is estimated to be 82 000 sq ft (7621 m^2);
(iii)   building costs are estimated to be £140 per sq ft [£1500 per m^2];
(iv)   finance costs are 14% pa;
(v)   the estimated building time is 18 months with construction starting in 6 months time;
(vi)   compensation to existing lessees together with planning and building regulation fees will amount to £150 000 and demolition costs are estimated at £100 000;
(vii)   a 3 months void period is to be allowed for letting after practical completion together with a rent free period of 3 months thereafter;
(viii)   estimated office rentals (today) are £50.00 per sq ft and there is considerable demand for offices of this size in this location.

Task B

Assuming that the site above is purchased for £22 775 725 (as shown in Table 2.3) calculate the profitability of the scheme.
The notes for Table 2.4 show also how the following returns are expressed:

Use feasibility method

(i)   profit yield
(ii)   annualised profit yield
(iii)   rental yield
(iv)   interest cover
(v)   rental cover
(vi)   equity yield

Refer to Table 2.4

Task C

Test how sensitive the yields in (B) above are to changes in rentals and yields using the data below:

Use sensitivity analysis

Rentals £45, £50, £55 per sq ft
Yields 6%, 6.25%, 6.50%

Refer to Table 2.5

Appraisal task and key data	Solution

Task D
Assuming (for illustration purposes) that all construction costs are paid quarterly in arrears in 6 equal stage payments calculate the following:

(i)　the revised profit for the scheme;
(ii)　the interest costs on a quarter by quarter basis;
(iii)　total interest charges.

Use cash flow method

Refer to Table 2.6

Task E
Assuming the same pattern of 'inflows' and 'outflows' over the development period calculate the following returns:

(i)　the net present value (NPV) of the scheme assuming a finance rate of 3.50% per quarter;
(ii)　the internal rate of return (IRR) on the same basis.

Note: A check can be carried out that shows the present value of the profit shown in Table 2.6 above equals the NPV.

Use discounted cashflow method

Refer to Table 2.7

Task F
Recalculate the residual land value upon the assumption that the developer has entered into a pre-funding agreement on the following terms:

(i)　the fund to receive a 'priority yield' of 6.25% on all costs incurred up to a maximum of £53 202 000;
(ii)　a rate of interest of 9% pa (2.25% per quarter) is to be charged by the fund on all monies expended including purchase of the site from the developer;
(iii)　if building is unlet the developer is to pay the fund the priority yield rental of £3 335 125 pa (monthly) but this liability to cease when the estimated profit of £10.64m is eroded;
(iv)　the developer to receive 50% of any rental income above £50.00 per sq ft capitalised at 6.25% (16 YP) subject to a limit on the average rental at £60.00 per sq ft.

Use forward sale method

Refer to Table 2.8

Task G
Assuming that the land owner has been advised that it would be preferable for him to retain the land and grant the developer a 125 years lease, calculate:

(i)　the minimum ground rental payable upon the assumption that the developer will pay £5m upon signing of the Building Agreement and a further £5m upon the full letting of the building; and
(ii)　show how the initial ground rental can be revised as and when the initial estimate of £50.00 per sq ft rental is exceeded either at first letting or at future rent review dates.

Use ground rent method

Refer to Table 2.9

Table 2.3 Appraisal of development land

				£
Gross Building Area:	Net	Rentals		
100 000 Sq ft:	82 000 sq ft	@£50 =		4 100 000
		Less Ground Rent (If Any)		
		Y.P. in Perpetuity @ 6.25 % =		16
		Gross Development Value		65 600 000
		Less Institutional Funding Fees/Legal		N/A
		Purchasers Costs =		
		Net Development Value		65 600 000

	Sub-Totals £	Sub-Totals £	
LESS			
A PRE-DEVELOPMENT COSTS			
1. Planning/Building Regs. Fees			
2. Compensation etc .... =	150 000	150 000	
B CONSTRUCTION COSTS			
1. Demolition	100 000		
2. Construction:			
100 000 sq ft @£140 per sq ft .... =	14 000 000		
3. Contingency .... @5 %	705 000		
4. Professional Fees .... @15 %	2 220 750		
5. VAT on Building .... @%	0		
6. VAT on Fees .... @%	0	17 025 750	
C INTEREST (Compounded Quarterly)			
1. Costs 24 Months .... @14 %	47 521		
2. Construction			
.... 50 % Costs 18 Months @14 %	1 951 620		
3. Void Period			
C1 .... 3 Months Interest @14 %	6 913		
C2 .... 3 Months Interest @14 %	664 208	2 670 262	
D LETTING/PROMOTION			
1. Agents Letting .... @15 %	615 000		
2. Agents Funding .... @1.5 %	984 000		
3. Legals on Funding .... @0.5 %	328 000		
4. VAT on Selling Fees .... @%	0		
5. Rent Free Period	1 025 000		
6. Promotion	50 000 =	3 002 000	
E PROFIT @ 20 % COSTS		10 935 000	33 783 012
F AVAILABLE FOR SITE COSTS		=	31 816 988
LESS 1. Interest on Total Site Costs		8 471 870	
2. Legals .... @0.5 %	113 879		
3. Surveyors Fees .... @1 %	227 757		
4. Stamp Duty .... @1 %	227 757 =	569 393 =	9 041 263
5. VAT on Acquisition Fees			
		RESIDUAL VALUE =	22 775 725

Notes for Table 2.3

Investment yield
The required investment yield is

$$\frac{4\,100\,000}{65\,600\,000} \times 100 = 6.25\,\%$$

Purchasers costs
Often an allowance of around 2.75 % of the investment valuation is allowed for costs of buying. In such circumstances the following calculation is necessary:

1.0275 × Net development value (NDV) = Gross development value (GDV)

$$\text{NDV} = \frac{\text{GDV}}{1.0275}$$

or in this example

$$\text{NDV} = \frac{£65\,600\,000}{1.0275}$$

$$\text{NDV} = £63\,844\,282$$

and therefore purchasing costs

$$= £63\,844\,282 \times 0.0275 = £1\,755\,713$$

However it is assumed for this example that as this is only an initial appraisal the developer may also be considering using different yields to assess the NDV and therefore purchasing costs can be disregarded for the purposes of this example.

Development costs
  (i) Includes all pre-development costs
 (ii) As supplied by the quantity surveyor
(iii) Contingency allowance. Usually 3−5 % of construction costs
(iv) Professional fees. Principal fees are:

    Architects
    Quantity surveyors
    Engineers (structural, services etc)
    Project management
    Funds supervision
 (v) VAT on building costs. Not included on buildings costs as VAT is assumed to be recoverable, i.e. developer to elect to charge VAT on rents or sale price of buildings. Note: Most computer programs allow now for interest on VAT for up to 3 months between payment of VAT inputs and VAT recovery.

Interest costs

(i) Over construction period of 18 months

Construction costs	£17 025 750 (REF C2 in Table 2.3)

Average at 50% over the building period	$\dfrac{0.5}{£8\,512\,875}$

multiply by:

Amount £1 at 3.5% for 6 quarters	=	1.229 2553
Total construction cost and interest	=	£10 464 497

Interest only	=	(£10 464 497 − £8 512 875)
	=	£1 951 620

Note: The amount of £1 $= (1 + i)^n$
where i = interest rate expressed as a decimal and n = quarterly interest payments (or annual, if applicable)

(ii) Over void period of 3 months
Interest here is calculated on total costs at the full interest rate of 3.50% per quarter:

Thus: Total costs	=	£17 025 750
plus interim finance	=	£1 951 620
Total costs to practical completion	=	£18 977 370
Interest for 3 months at 3.5%	=	£664 208

Allowance for profit and risk
Based on total cost of land and buildings, but if land cost not known and if the profit is 20% on costs then the profit becomes:

NDV	=	£65 600 000
divided by		1.20
Therefore NDV (less profit)		£54 666 667
and profit	=	£10 935 000 (say)

The residual land value (RLV)
This is then worked out and purchasing costs and interest calculated.

If interest on land costs is taken at 3.5% per quarter for 9 periods and land purchasing costs are 2.5% of the RLV then the formula below will calculate the RLV.

$(1.035)^9 \times 1.025$ RLV   =   £31 816 988

1.3628974×1.025 RLV   =   £31 816 988

      RLV   =   £31 816 988

                1.39696973

and the RLV   =   £22 775 729

Calculation of interest costs on land purchase
RLV (as above)   £22 775 729

Add costs of purchase at 2.5%   =   £   569 393

Total land acquisition costs   =   £23 345 122

Then multiply by (1.035) to find total interest costs after land purchasing costs.
Amount of £1 at 3.5% per quarter for 9 quarters   =   1.362897

Total available for site costs   =   £31 816 990

i.e. interest on site purchasing costs (say)   =   £8 471 870

Thus it can be seen that the residual land value is (say)   £22 775 725

Table 2.4 Feasibility analysis

	Net	Rentals		£
Gross Building Area: 100 000 sq ft:	82 000 sq ft	@£50 =		4 100 000
		Less Ground Rent (If Any)		
		Y.P. in Perpetuity @6.25% =		16
		Gross Development Value		65 600 000
		Less Institutional Funding Fees/Legal		N/A
		Purchasers Costs =		
		Net Development Value		65 600 000

	Sub-Totals £	Sub-Totals £
LESS		
A PRE-DEVELOPMENT COSTS		
1. Planning/Building Regs. Fees		
2. Compensation etc	150 000	150 000
B CONSTRUCTION COSTS		
1. Demolition	100 000	
2. Construction:		
100 000 sq ft @£140 per sq ft .... =	14 000 000	
3. Contingency .... @5%	705 000	
4. Professional Fees .... @15%	2 220 750	
5. VAT on Building .... @%	0	
6. VAT on Fees .... @%	0	17 025 750

Table 2.4 contd.

C	INTEREST (Compounded Quarterly)		
	(1). Costs 24 Months .... @14%	47 521	
	(2). Construction		
	.... 50% Costs 18 Months @14%	1 951 620	
	(3). Void Period		
	C(1) .... 3 Months Interest @14%	6 913	
	C(2) .... 3 Months Interest @14%	664 208	2 670 262
D	LETTING/PROMOTION		
	1. Agents Letting .... @15%	615 000	
	2. Agents Funding .... @1.5%	984 000	
	3. Legals on Funding .... @0.5%	328 000	
	4. VAT on Selling Fees .... @%	0	
	5. Rent Free Period	1 025 000	
	6. Promotion	50 000	3 002 000
E	SITE COSTS		
	1. Land Price (From Table 2.3)	....	22 775 725
	2. Interest on Total Site Costs	....	8 471 870
	3. Legals .... @0.5%	113 879	
	4. Surveyors Fees .... @1%	227 757	
	5. Stamp Duty .... @1%	227 757	569 393
	6. VAT on Acquisition Fees		

	=	54 665 000
RESIDUAL PROFIT	=	10 935 000

Notes for Table 2.4

The land price is now assumed to be known and the profit is worked out as a residual profit. Important yields are as follows:

(i) Profit yield
This is the profit as a percentage of costs.

$$\text{A} \quad \text{Profit} = £10\,935\,000$$
$$\text{B} \quad \text{Costs} = £54\,665\,000$$

Therefore yield on cost $= \dfrac{\text{A}}{\text{B}} \times 100 = 20\%$

(ii) Profit yield annualised
The above shows the profit as a percentage of total costs over the entire period of the development. The annualised profit yield is calculated by compounding back the profit yield figure on an annual basis, i.e.:

A Yield on cost = 20%
B Development period 27 months (2.25 years)
   Annualised yield $= {}^{2.25}\!\sqrt{20} = 8.44$

(iii)   Rental yield
First year's net rent as a percentage of cost.

$$A \quad Rent = £4\,100\,000$$
$$B \quad Costs = £54\,665\,000$$

Therefore rental yield $= \dfrac{A}{B} \times 100 = 7.50\%$

(iv)   Interest cover
The period of time from the end of the void period until the profit is eroded by the interest charges assuming that the interest rate in the appraisal remains constant.

$$A \quad Profit = £10\,935\,000$$
$$B \quad Interest\ at\ 14\%\ on\ £54\,665\,100 = £7\,653\,100\ pa$$

Interest cover $= \dfrac{A}{B} = 1.43$ years

(v)   Rental cover
This is the period of time from the end of the void period until the profit is *eroded* by the equivalent rents payable assuming that the rent payable remains constant.

$$A \quad Profit = £10\,935\,000$$
$$B \quad Rental = £\ 4\,100\,000$$

Rental cover $= \dfrac{B}{A} = 2.67$ years

(vi)   Equity yield (not used in this analysis)
This type of yield calculation is related to the developer's own return on the capital he puts into the project — the yield will depend particularly upon:

(a) the ratio of loan: total costs;
(b) the rate of interest on borrowed capital;
(c) the future rental income from the scheme;
(d) the yield on the investment;
(e) safeguards used to protect the risks involved.

Not all the yields referred to in (i)–(vi) are used simultaneously in practice. The emphasis given to each will depend primarily upon the financial status of the developer undertaking the scheme and the method(s) of funding being used for the project. For example, if the developer is entering into a forward funding transaction with 100 % pre-funding arranged then he will give more emphasis perhaps to (i), (iv) and (v) whereas an 'investing developer' using a large part of his own capital resources will give particular emphasis to (ii) and (vi) as the short term funding profit is of less interest to him than the long term worth of the development as an investment

Table 2.5 Sensitivity analysis

RENT £45.00	Yields 6.00 %	Yield 6.25 %	Yield 6.50 %
NDV £	61 500 000	59 040 000	56 769 232
COSTS £	54 520 268	54 471 068	54 425 652
PROFIT £	6 979 732	4 568 932	2 343 580
INTEREST COVER (yrs)	0 yrs 11 mths	0 yrs 7 mths	0 yrs 3 mths
RENTAL YIELD %	6.77	6.77	6.78
PROFIT YIELD %	12.80	8.39	4.31
(Annualised %)	5.50	3.64	1.89
RENT £50.00			
NDV £	68 333 336	65 600 000	63 076 924
COSTS £	54 718 436	54 663 772	54 613 308
PROFIT £	13 614 900	10 935 000	8 463 616
INTEREST COVER (yrs)	1 yr 8 mths	1 yr 4 mths	1 yr 1 mth
RENTAL YIELD %	7.49	7.50	7.51
PROFIT YIELD %	24.88	20.01	15.50
(Annualised %)	10.38	8.44	6.61
RENT £55.00			
NDV £	75 166 664	72 160 000	69 384 616
COSTS £	54 916 604	54 856 468	54 800 964
PROFIT £	20 250 060	17 303 532	14 583 652
INTEREST COVER (yrs)	2 yrs 4 mths	2 yrs 1 mth	1 yr 9 mths
RENTAL YIELD %	8.21	8.22	8.23
PROFIT YIELD %	36.87	31.54	26.61
(Annualised %)	14.97	12.96	11.06

Notes for Table 2.5

For simplicity one example only of sensitivity analysis will be illustrated here.

From the table it can be observed that a rental of £50.00 per sq ft combined with a yield of 6.25 % will give a profit of £10 935 000 but if the investment yield falls to 6.00 % and the rent rises to £55.00 then the profit nearly doubles to £20 250 060. On the other hand for a situation where the rental is £45.00 per sq ft and the yield is 6.50 % a profit of only £2 343 580 will be produced (i.e. only 21 % of the originally estimated profit for the scheme).

Table 2.6 Cash flow statement

QTR	OUTFLOWS £	INFLOWS £	BAL BEFORE INTEREST £	INTEREST @ 3.5% £	CF BALANCES £
0 Land	23 345 118		23 345 118	817 079	24 162 197
1 Possession	150 000		24 312 197	850 927	25 163 124
2	2 837 625		25 163 124	880 709	26 043 833
3	2 837 625		28 881 458	1 010 851	29 892 309
4 Construction	2 837 625		32 729 934	1 145 547	33 875 482
5 over	2 837 625		36 713 107	1 284 958	37 998 066
6 six	2 837 625		40 835 691	1 429 249	42 264 940
7 quarters	2 837 625		45 102 565	1 578 589	46 681 155
8	2 837 625		49 518 780	1 733 157	51 251 937
9 Letting etc	3 002 000	65 600 000	11 346 063	SAY	11 346 000
SUB TOTALS	43 522 863	65 600 000		10 731 066	11 346 000

CHECK

		INFLOW =	65 600 000
LESS (1)	Outflows	43 522 863	
(2)	Interest	10 731 066	= 54 253 929
		PROFIT =	11 346 071
		(as above)	
		SAY	11 346 000

Notes for Table 2.6

Table 2.6 shows the interest that will be incurred on the project. The final surplus on the cash flow is £11 346 000 made up as follows:

Total land and building costs	£43 522 863
Add interest as per cash flow	£10 731 066 (total interest)
Total costs	£54 253 929
Total receipts	£65 600 000
Estimated profit at end of project (say)	£11 346 000

Note: This profit is higher than the original estimate on the residual valuation (Table 2.3) but again is only an estimate since rents, building costs and so forth, will undoubtebly change over the 27 months allowed for the scheme.

Table 2.7 Discounted cash flow statement

QTR	OUTFLOW £	NOTES	INFLOWS £	PV 3.5%	(−) DCF £	(+) DCF £
0	23 345 118	Land		1	23 345 118	
1	150 000	Possession etc		0.966	144 927	
3	2 837 625			0.902	2 559 375	
4	2 837 625	Construction		0.871	2 472 826	
5	2 837 625	over		0.842	2 389 204	
6	2 837 625	six		0.813	2 308 409	
7	2 837 625	quarters		0.786	2 230 373	
8	2 837 625			0.759	2 153 757	
9	3 002 000	Letting etc.		0.73373	2 202 660	
9		NDV	65 600 000	0.73373	=	48 132 688
TOTALS	43 522 868		65 600 000			

	TOTALS	39 806 649 48 132 688
	NPV =	8 326 000
	INTERNAL RATE OF RETURN = [Where NPV = 0]	28%

Notes for Table 2.7

Net present value (NPV)
The NPV is the difference between the present values of the inflow and outflows when the target rate = 3.5% per quarter, i.e. the NPV = £8 326 000.

The NPV is in fact the present value of the future profit shown in Table 2.6 (cash flow). It shows the project is viable and as a check the following calculations can be made:

Profit in 27 months	=	£11 346 063
PV £1 for 27 months at 3.5 % per quarter	=	0.73373
Net present value (say)	=	£ 8 326 000

Internal rate of return (IRR)
Of wider application in development appraisal, is the calculation of the IRR — i.e. the breakeven yield, when neither a profit or loss is incurred. The difference between the discount funding rate of 14 % and IRR represents the 'profit and risk' element and so the interest rate would have to double (from the start) to completely wipe out the profit.

The IRR is found by a process of iteration by computer or commonly by trial and error using various target rates to obtain a negative and positive NPV — thus a trial rate of say, 8 % per quarter produces a negative (loss) situation and a trial rate of say 6.5 % per quarter, a positive position. Interpolation between these figures produces an IRR of 28 % (annualised), which normally would be considered to be a very acceptable return on this project especially as the return may be enhanced by several factors such as careful project management of the scheme, by an improvement of funding terms or by market rentals moving upwards during the development period.

Table 2.8   Pre-funding example

Appraisal of development land				FORWARD SALE £
Gross Building Area:	Net	Rentals		
100 000 Sq ft:	82 000 sq ft	@£50	=	4 100 000
		Less Ground Rent (If Any)		
		Y.P. in Perpetuity @ 6.25 %	=	16
		Gross Development Value		65 600 000
	Less Institutional Funding Fees @ 2.75 %			1 755 720
		Purchasers Costs	=	
		Net Development Value		63 844 280

	Sub-Totals £	Sub-Totals £
LESS		
A PRE-DEVELOPMENT COSTS		
1. Compensation of Tenants		
2. Site Clearance	150 000	150 000
B CONSTRUCTION COSTS		
1 Demolition	100 000	
2 Construction:		
100 000 Sq ft @ £140 per Sq ft =	14 000 000	

Table 2.8 Contd.

Appraisal of development land			FORWARD SALE
3. Contingency .... @5 %	705 000		
4. Professional Fees .... @15 %	2 220 750		
5. VAT on Building .... @%	0		
6. VAT on Fees .... @%	0	17 025 750	
C INTEREST (Compounded Quarterly)			
(1). Costs 24 Months .... @9 %	29 225		
(2). Construction			
.... @50 % Costs 18 Months @9 %	1 215 858		
(3). Void Period			
C(1) .... 3 Months Interest @9 %	4 033		
C(2) .... 3 Months Interest @9 %	410 436	1 659 552	
D LETTING/PROMOTION			
1. Agents Letting .... @15 %	615 000		
2. Agents Funding .... @1.5 %	984 000		
3. Legals on Funding .... @0.5 %	328 000		
4. VAT on Selling Fees .... @%	0		
5. Rent Free Period	1 025 000		
6. Promotion	50 000 =	3 002 000	
E PROFIT @ 20 % COSTS		10 642 281	32 479 581
F AVAILABLE FOR SITE COSTS		=	31 364 697
LESS 1. Interest on Total Site Costs		5 692 014	
2. Legals .... @0.5 %	125 233		
3. Surveyors Fees .... @1 %	250 465		
4. Stamp Duty .... @1 %	250 465 =	626 163 =	6 318 177
5. VAT on Acquisition Fees			
	RESIDUAL VALUE	=	25 046 520

Notes for Table 2.8

The developer is to receive the following payments:

(i) The pre-arranged buyout profit of 16 years purchase (YP) of the rental achieved above the priority rental (less the funds costs) as follows:

Estimated rental at £50 per sq ft (if achieved)		=	£4 100 000
Less priority yield of 6.25 %			£3 325 125
(on original cost or as adjusted)			
	Overage		£774 875
	YP		16
			£12 398 000
Less fund's purchasing and funding costs			£1 755 720
at 2.75 % NDV			
Net payment to the developer (say)		=	£10 642 000

(ii) Plus a further payment as follows:

Developer to receive 50% share of rental up to £60.00 per sq ft

i.e. 82 000 Sq ft × £5.00 per sq ft	=	£410 000
YP (agreed)	=	15
		———
Additional payment to developer (assume net of costs)		£6 150 000

   The fund's return can be expressed as both an initial yield (i) and a profit yield (ii) calculated as follows:

(i)   Initial yield
The initial yield to the fund can be measured by the return on costs compared with its target priority yield of 6.25%.

Fund's total cost (as estimate)	£53 202 000
Add 2 payments to developer (as above)	£16 792 000
	———
Total costs =	£69 994 000

If final rental is £65.00 per sq ft then the final rental yield =

$$\frac{£5\,330\,000}{£69\,994\,000} \times 100 = 7.6\%$$

(ii)   Profit yield

Final rental = £65.00 per sq ft	=	£ 5 330 000
YP in perpetuity at 6.25%	=	16
		———
Gross development value	=	£ 85 280 000
Less costs (as above)		£ 69 994 000
		———
Profit	=	£15 286 000

Profit yield to the fund $= \dfrac{£15\,286\,000}{£69\,994\,000} \times 100 = 21.839\%$

Table 2.9 Ground rental valuation

A			£	£
Summary	A	PRE-DEVELOPMENT COSTS		150 000
of Costs	B	CONSTRUCTION COSTS		17 025 750
(as Table 2.3)	C	INTEREST ON (A) + (B)		2 670 262
	D	LETTING/PROMOTION/FINANCE		3 002 000
	E	PREMIUM (First)	5 000 000	
		Interest for 9 QTRS @ 3.5%	1 814 486	

Table 2.9 Contd.

	PREMIUM (on letting)	$\underline{5\,000\,000}$	= 11 814 486
	F LEGAL COSTS		100 000
	(Building agreement, ground lease etc.)		
		TOTAL COSTS	34 762 498
B Minimum Annual Returns to Fund and Developer	a) FUND'S YIELD at 7.50 % of costs = b) Developer's Profit at 2 % of costs =		2 607 187 695 250
		i.e. 9.5 % of costs =	3 302 437
C Calculation of Ground Rental	ESTIMATED NET INCOME *LESS* as above THEREFORE AVAILABLE FOR GROUND RENTAL SAY, *Summary* a) Ground rent as % of net income = b) Return to fund as % of net income = c) Return to developer as % of net income =		4 100 000 3 302 437 797 550 pa 19.45 % 63.59 % 16.96 %
			100.00 %
D Revision of Ground Rental on Letting of the Building	IF BUILDING LETS AT £5 330 000 pa (£65.00 psf) i.e. £1 230 000 pa income above estimates, the revised GROUND RENTAL becomes:	Minimum g.r. = *add* 19.45 % of £ 1 230 000 =	797 550 239 235
		Therefore Revised G.R =	£1 036 785
E Recommended Revision of Ground Rental on Review	The 5 years REVIEWED g.r. thereafter is 19.45 % the OCCUPATIONAL RENTS RECEIVABLE FROM LESSEES		

Notes for Table 2.9

(i) Investment yield
This is taken at 0.75 % above the equivalent freehold investment yield to reflect the fact that a 125 years lease is less attractive than for an equivalent freehold investment of the same type of investment.

(ii) Developer's profit
The profit represents 21.4 % of the investment yield in this example but will of course vary with each developer and his desire to succeed in his bid for the site under consideration.

(iii)   Annual sinking fund
Sometimes an allowance to replace the building after a certain number of years is incorporated into the calculation but in this example it is incorporated in (ii) above.

(iv)   Increases in building costs, etc.
If costs rise, the returns in (i) and (ii) will also increase although (ii) may well be 'capped' at the original estimated return to the developer.

(v)   The revised ground rental
The percentage payable to the landowner of any overage resulting from a better than anticipated rental will be arrived at by negotiation and in this situation it is assumed to be at the same ratio as previously assessed — a much higher percentage than this may cause the ground rental to rise to a level at which the investment yield is affected by the greater risk of a high ground rental liability.

IMPLICATIONS ON QUANTITY SURVEYING ADVICE AND PRACTICE

Quantity surveyors' input to development appraisal will involve their recognised skills in various aspects related to construction costs, procurement, programming and so forth. Well developed techniques and procedures exist for producing such advice and it is not the purpose of this chapter to cover these aspects.

Instead some of the special requirements placed upon the quantity surveyor's input when involved in development appraisal will be discussed. These are summarised under the following headings:

- Building and construction costs
- Professional fees
- Cashflow
- Risk
- Sensitivity
- Forecasting values and costs
- Floor areas
- Mixed development schemes — apportionments
- Partnership schemes
- Programme
- Procurement
- Planning gains/agreements under the Town and Country Planning Acts
- Value added tax

Building and construction costs

This is the quantity surveyor's central role in development appraisal. In turn, building and construction costs will be in most cases the major element of development costs and will have a critical bearing on the profitability of the development. Since the principal objective of the developer is to earn a profit or return, the quantity surveyor's advice in assessing these costs will need to be tailored to the special needs of the developer and the appraisal team. Major aspects to be considered will be:

(a) Site development costs, noting which of these might be used in land price negotiations.
(b) The building elements of the 'developers provision or works'.
(c) Demarcation between developers provision and reasonable tenants/occupiers fitting out.
(d) Associated on-costs including fees.

Also it should be borne in mind that the developer or the appraisal team will use the base estimate for building costs when generating other 'on-costs' such as VAT, building finance (interest), the cost of delays, building finance for void lets and so forth.

The various methods of estimating developed by the quantity surveying profession are used quite extensively in development appraisal. Initial appraisals will by necessity, due to the lack of definition of the project, be based on the quantity surveyor's costs per square metre approach or a program such as the RICS 'ELSIE' expert system. As the scheme develops and funders are sought for commitment to the scheme an approximate quantities or adjusted elemental cost plan approach will be the more appropriate. Prior to construction on site a pre-tender estimate can be made available. However, this is not commonly required by developers who by this stage normally prefer to wait until tenders are received.

Professional fees

At the initial appraisal stage the developer will include an appropriate allowance for professional fees associated with design and implementation which will be calculated on the total of the building estimate. Very often the developer will base this allowance on experience gained from securing commissions on previous schemes. However, eventually this will be supported by a firmer view on the likely fee level for design consultants. Some developers seek advice from quantity surveyors on fees since they are usually in a better position than most to assess the various component values of a scheme upon which fees are based.

There can be no hard and fast rules as to the extent to which quantity surveyors should advise developers on their provision for professional fees. This will depend on the quantity surveyor's own knowledge of a particular project and the project implementation practices of the particular developer. Whatever the extent of his advice the quantity surveyor should be aware of factors within the scheme that can distort the conventional allowances for professional fees. These might include a higher than normal civil engineering content, the necessity to obtain inputs for traffic management, the sensitivity of environmental conditions which might necessitate the appointment of specialist advisory inputs and so forth.

Cashflow — timing of payments

For ease of illustration, the cash flow example contained in Table 2.6 in the previous section has been prepared on the assumption that construction costs will be expended in equal amounts per quarter. In practice, where requested by the developer, the quantity surveyor will be required to predict the actual pattern of cash flow and prepare estimates of the sums involved.

The normal distribution for construction cash flows is the 'lazy S' but on many projects this pattern will not be achieved. The quantity surveyor will need to analyse the scheme in order to identify factors that will distort the normal distribution. Schemes with high mobilisation costs (e.g. site facilities on say Central London refurbishment sites) or complex preparation works will mean 'front-loaded' schemes and hence the potential for higher interest costs for building finance. By contrast, schemes with higher than normal fitting out content will produce back-loaded costs and interest for building finance will be lower.

Risk

Uncertainty (or risk) is inherent in all development but the property professions have not adopted in any comprehensive sense methods of quantifying the effects of risks on costs and presenting the results in terms of their probability. Additionally it might be considered somewhat pointless for the quantity surveyor to present construction costs in terms of probability if the developer and the rest of the team are using deterministic figures.

However for complex schemes involving relatively long development periods, developers and investors are demanding profit or returns to be expressed in terms of their probability. This involves the appraisal team in identifying what risks are likely to be encountered on the project. Next the effects of these risks on the various cost and income components are estimated and thirdly computer aided simulations are carried out which attempt to combine the various cost and income components and measure the overall probability of financial targets being achieved.

Techniques and presentations such as decision trees, Monte-Carlo simulation and utility analysis are available to the quantity surveyor as an aid to the assessment of the risk contained in a development.

Sensitivity

Many development appraisals involve testing the underlying assumptions and estimates with some form of sensitivity analysis. Although it is recognised that interest rates, yields and rental levels (the latter with its dependency on the interaction of demand and supply) are the more volatile components within the appraisal, it is acknowledged that at certain times building costs can also be subject to extremes of volatility. The quantity surveyor may therefore be required to present his cost in terms of a range of figures.

The range could reflect the effects of possible changes in market conditions (e.g. tender price inflation), scheme variations (especially where a range of specifications or infrastructure works are under consideration) or any other significant variable (e.g. construction periods).

Forecasting values and costs

It is essential that the quantity surveyor seeks guidance from the developer or the appraisal team as to the basis of all estimates and forecasts. It may seem an obvious (and hopefully, nowadays, redundant) point to make that if rentals and other income related variables are based on current non-inflated rates distortions in the appraisal will occur if costs are assessed on a different (out-turn) basis.

Some developers may take the view that rentals and other incomes are the more difficult to predict several years in advance whereas building costs are the more constant and hence greater distortions are possible if rentals are inflated. This might be held to be a fair comment but the quantity surveyor should encourage others contributing to the appraisal to harmonise the basis of all estimates.

Even if an initial residual appraisal is produced by assessing all figures on a common basis, this may not be the case later in the scheme. For example, once a scheme has passed to a more detailed stage and a feasibility or a cashflow approach is adopted then the developer will wish to deal with best estimates of actual expenditure and insist upon out-turn prices. At this stage the developer (or his agent) may take the view that rentals should be included at their current market expectation rather than projected.

Floor areas

Another matter to be clarified at the earliest possible stage in any development appraisal is the measurement and application of floor areas.

Quantity surveyors normally present their cost data in terms of gross building areas measured to the inner face of the external walls. Rentals generally are applied to net lettable areas which can be anything from 10–20% less than the quantity surveyor's areas. In cases where the developer intends to sell the freehold of the development the quantity surveyor's area may be the more appropriate to use although in some cases gross external building areas may be adopted (i.e. measured to outer face of external walls). Architects and other designers may need to generate floor area data (for example in calculating certain planning densities or plot ratio) which may differ again from those of the quantity surveyor and developer.

This aspect becomes the more critical on developments where several forms of completed disposals are involved. It is essential that all members of the appraisal team clarify the basis of floor area measurements and the use to which they will be put.

Mixed development schemes/apportionments

Many development schemes comprise a variety of uses which also attract a variety of funding/completed disposal methods. Even a development with a predominant single use such as retail may involve part of the scheme being sold to a major anchor store on a freehold basis, with the balance held as an investment by the developer and funded in a conventional manner. In addition, above the retail there may be small residential units for sale.

Developments of business space can vary in their content and methods of disposal. Parts may be retained by the developer, other parts may be sold or let to an owner occupier.

In these situations the quantity surveyor will be asked to analyse his estimates so that the developer is able to assess the profitability of each component and clear guidance will need to be obtained on the allocation of external works, foundations, party floors and walls and so forth.

Also, the developer may need to calculate service charges on a discriminating basis for different parts of the development. Certain costs contained within the quantity surveyor's estimate may be used in such a calculation.

Partnership schemes

Quantity surveyors are involved in partnership schemes in a variety of ways. Most commonly they can be found to be acting directly for the developer in a conventional quantity surveying role or in a watching brief/development monitoring role on behalf of local authorities, funders and so forth.

Irrespective of a quantity surveyor's particular role there needs to be a common understanding of the components of a partnership and its effect on data including:

(a) The terms of the financial framework within the partnership agreement.
(b) The conditions of the building agreement with special emphasis on:

- specifications;
- drawings;
- extent of built facilities;
- constraints imposed upon the developer;
- sequencing; and
- other special conditions arising from the partnership which would not normally be anticipated from single party developments.

Programme

Any appraisal of a development opportunity will need to incorporate a view on the required development period. This will be essential in order to assess the costs of financing the scheme and, in terms of predicting demand, when the development will be completed.

However the developer will have objectives other than time involving quality (specification) and cost (budget). The appraisal team will need to strike a balance amongst these objectives and this will involve the quantity surveyor in assessing the implications of maintaining such a balance.

For example, overlapping design and construction to such an extent that expenditure on the building site becomes inefficient may not benefit the scheme as much as if a slower pre-construction period had been developed with a correspondingly faster construction phase.

Although a start on site is recognised as a key milestone in any development, developers will take differing views as to the costs (risks) they are prepared to underwrite at any particular stage for commissioning the professional effort required to progress a scheme to commencement on site. Hence, should a lengthy planning process be envisaged and a developer is unwilling to commission the team until consent is obtained, acceleration of the design and tendering process may need to be considered if the start on site date becomes critical.

If the quantity surveyor is asked to assess or advise on the development programme it will be necessary to obtain the maximum briefing on all aspects having an effect on time. This will involve not only matters with which the quantity surveyor is directly concerned (e.g. periods for design, mobilisation, construction, commissioning and testing, handover and so forth) but also other matters which will need to be advised upon by other members of the appraisal team. These matters will include the timing of planning applications and of legal agreements for site exchange, the possibility of a compulsory purchase enquiry and so forth. Also, the quantity surveyor should discuss how realistic or immoveable are any preferred completion dates. Often developers attempt to provide buildings for certain key times of the year (e.g. a Christmas opening for retail developments). In reality allowing for proper fitting out periods can sometimes mean these targets are not achievable and the alternative preferred completion dates can entail a significant extension to the available development period.

Procurement

Very much linked to some of the aspects discussed under programme, procurement choices can exert a significant influence on the quantity surveyor's estimates. Many attempts have been made to provide guidance and classifications of the advantages, disadvantages and the risks associated with the various recognised procurement routes. It is beyond the scope of this chapter to discuss these various aspects but the quantity surveyor should make the developer and team aware of the implications of electing to adopt any particular procurement method.

Planning gain/agreements under the Town and Country Planning Acts

Increasingly part of the return earned from developments is required to be used for works or benefits which are commonly classified as 'planning gain'. Usually commitment from the developer to provide such benefits is obtained by an agreement made in accordance with section 52 of the Town and Country Planning Acts.

The development costs associated with some planning gain can be significant and quantity surveyors must clarify the extent to which such items should be included within their estimates.

It is beyond the scope of this chapter to illustrate the various ways in which planning gain can be provided but it is important to note that the gain may be provided through built facilities located on other sites and provided not necessarily within the time frame of the enabling development.

Value Added Tax

Regulations in connection with Value Added Tax have simplified its application to construction work. For the most part all construction work and associated fees involved in development projects will be subject to a positive rate of VAT. The treatment of VAT within a development appraisal will depend upon the tax status of the developer and the corporate and financial arrangements to be adopted for the scheme including the ability or desire to exercise the election to tax option.

Quantity surveyors will need to obtain clear instructions as to the extent to which they are to include or exclude VAT when preparing cost advice in connection with development appraisals.

BIBLIOGRAPHY

Principles of valuation

Britton, W., Davies, K. and Johnson, T. (1989) *Modern Methods of Valuation*, 8th edn. London: The Estates Gazette Ltd.

Darlow, C. (ed.) (1982) *Valuation and Investment Appraisal*, 7th edn. London: The Estates Gazette Ltd.

Darlow, C. (ed.) (1988) *Valuations and Development Appraisal*, 2nd edn. London: The Estates Gazette Ltd.

Davidson, A. W. (ed.) (Published annually) *Parry's Valuation Tables.* London: The Estates Gazette Ltd.

Enever, N. (1989) *Valuation of Property Investments.* London: The Estates Gazette Ltd.

Millington, A. (1982) *Introduction to Property Valuation*, 2nd edn. London: The Estates Gazette Ltd.

Rees, W. (ed.) (1988) *Valuation: Principles into Practice*, 3rd edn. London: The Estates Gazette Ltd.

Richmond, D. (1985) *Introduction to Valuation*, 2nd edn. London: Macmillan.

Development

Bows, D. (1983) *A checklist for Development and Finance Agreements*. Derby: HS Publications.

Byrne, P. Cadman, D. (1984) *Risk, Uncertainty on Decision in Property Development*. London: E. & F.N. Spon Ltd.

Cadman, D. and Austin Crowe, L. (1983) *Property Development*, 2nd edn. London: The Estates Gazette Ltd.

Marber, Paul and Marber, Paula (1985) *Office Development*. London: The Estates Gazette Ltd.

Martin, P. (1982/3) *Shopping Centre Management*. London: E. & F.N. Spon Ltd.

Information technology

Kirkwood, J. (1984) *Information Technology and Land Administration*. London: The Estates Gazette Ltd.

Property management and report writing

Torkildsen, G. (1986) *Leisure and Recreation Management*. London: E. & F.N. Spon Ltd.

Investing

Barter, S. (1988) *Real Estate Finance*. Sevenoaks: Butterworth.

Marshall, P. J. L. *Donaldsons Investment Tables*, 3rd edn. London: Surveyors Publications, RICS.

Yates, A. and Gilbert, B. (1989) *The Appraisal of Capital Investment in Property*. London: Surveyors Publications, RICS.

Chapter 3

Value Management

JOHN KELLY, *Heriot-Watt University, Edinburgh* and
RUSSELL POYNTER-BROWN, *Dearle and
Henderson, London*

INTRODUCTION

This chapter is concerned with the techniques currently used by value engineers in North America in the provision of a service to construction industry clients which attempts to provide the required quality of building at the least cost.

Value management is a philosophy and a set of techniques which are each addressed in this chapter. The philosophy centres on the identification of the function of a space or a component. The techniques rely on a method termed the 'job plan' which is explained in the context of its application at the concept, design and construction phases of a project. The formal approaches applied in North America are outlined together with the subjects of design liability and the relationship of value management to project management. Finally, its compatibility with current quantity surveying practice and the opportunities for quantity surveyors are discussed.

The background information for this chapter was obtained during research undertaken by John Kelly and Steven Male of the Department of Building, Heriot-Watt University. The RICS Education Trust provided a travel grant and the project was managed by Russell Poynter-Brown, Brian Gilbert with Keith Hudson as chairman under direction from the RICS QS R & D Committee.

BACKGROUND

Value management is a philosophy concerned with providing the product desired by a customer at the required quality and the optimum cost. The philosophy evolved during the 1940s when the manufacturing industry in the USA was under pressure to supply goods made from scarce materials and components. The General Electric Company (GEC) developed a technique whereby such problems were overcome by analysing the function of the unavailable component and resourcing the function by alternative design and/or component combinations. The technique was successful because it concentrated on the function of the parts rather than on their shape or material qualities. It was soon realised that some of the alternatives were less expensive than the original design and yet were able to provide the same function at an equal quality.

The realisation that equal or better performance could be provided at lower cost led GEC to set up a process of value analysis following what became the first definition of value management which is:

an organised approach to providing the necessary functions at the lowest cost.

This approach, which was applied to design was also applied to products which were well resourced and which were currently passing down the production lines. The philosophy here was that all products contain unnecessary cost. This presumption led to a second definition of value management which is:

an organised approach to the identification and elimination of unnecessary cost where unnecessary cost is that which provides neither use, life, quality, appearance, nor customer features.

In respect of building and within this definition:

- Use refers to the utility of the component, space within a building, etc. This utility is measured by reference to the extent to which it fulfils the required function.
- The life of the component or material must be in balance with the life of the whole building. For example, unnecessary cost may be introduced if a component is specified which has a life of 60 years within a building which has a predicted life of 15 years. Similarly a component which has a design life of 15 years will introduce unnecessary cost if the life of the building is to be 20 years, since its replacement will then be out of balance.
- Quality is a subjective function, but however it is perceived it must be preserved. The philosophy of value management looks towards reducing cost without sacrificing quality.
- To most architects, interior designers and landscape architects, the appearance of the building is its most important attribute. The form of the building obviously follows its function and therefore allows the building to be utilised but designers will arrange the spaces and configure the components in a way which is sensitive to aesthetics. Engineers, be they structural, mechanical or electrical, have always respected the design in reaching their solution and it is the responsibility of those practising value management to do the same. However, those components and spaces which do not contribute to appearance require analysis.
- Customer features are those which sell, the graphics often applied to manufactured goods are an example. In the context of building, examples of customer features may be found in interior design and shop fitting and sometimes in the shape of the building itself. The client may wish to express the corporate image within the building or by the building itself and this aspect should not be subject to deletion by value management.

It may appear that the above interpretation of the two definitions leaves little scope for the provision of the necessary functions at the lowest cost or for the identification and elimination of unnecessary cost. However, most value managers in North America work on the principle that at least 10 % of the estimated project cost can be saved by the application of these two simple definitions.

It should be noted that in North America those who carry out value analysis are termed value engineers. In the UK the terms value management and value managers have been adopted.

THE JOB PLAN

All value management techniques adopt a series of operational steps which have become known as the 'job plan'. Alternative names for the steps have been used but they can all be summarised under the following five headings:

Information
Creativity
Judgement
Development
Recommendation

Information

At this stage the spaces, elements and components of the proposed building are identified in terms of function. The understanding of function allows an appreciation of alternatives to a standard solution and is normally triggered by the question 'what does it do?'. This approach to information gathering can also confirm the client's intentions. For example, during one value management exercise conducted on a prison project in North America, it became apparent that the design team saw the function of the cell door as being, 'to keep the prisoner in', whereas the client's team knew that it was, 'to keep other prisoners out'. This simple mismatch of concept was corrected by asking 'what does it do?'.

Information regarding the function of spaces within the building can also facilitate an understanding of client requirements, for example two parts of the client organisation may specify the same space by a different name resulting in duplication in the design. Similarly in one building the height of a space was specified by the future user as being 1 metre higher than the tallest piece of equipment to be accommodated in that space. The design team took the stated height to be a minimum and allowed a further 1 metre contingency, under the 1.2 metre deep ventilation ducting which passed through the same space. This resulted in the ceiling being 3.2 metres above the tallest piece of equipment.

The gathering of information based upon function is important because it allows the full purpose and consequences to be realised. In any decision making process the quality of the decision cannot rise above the quality of the data upon which the decision was based.

In laying the foundation for later generation of ideas and decision making it is essential that the expression of the function is made in as few words as possible, normally by a noun plus a verb. For example the function of a window, in its normal form as a sheet of glass in an opening casement, might be:

Provide light
Permit ventilation
Prevent access
Retain heat
Insulate from external noise
Permit view

Additionally there may be functions which may not be required and may necessitate expenditure to alleviate their results, e.g.

Allow solar heat gain
Allow distracting sunlight (glare)
Generate cold radiation

As a part of the information process the functions of the subject under review are categorised as either primary or secondary functions. For example consider a window in a deep plan office which is designed to be artificially lit and ventilated. In this case the primary functions are:

Prevent access
Retain heat
Insulate from external noise
Permit view

The other functions, (provide light and permit ventilation), become secondary since these are provided artificially as a result of the deep plan configuration. The other functions of allow solar heat gain, allow distracting sunlight (glare), and generate cold radiation are also secondary since they are unwanted results of the technological solution.

This logical analysis of the function of the space, or the 'normal' technological solution for an element or component sheds a new light on the information given by the client organisation or on the first solution generated by the design team. The division of the functions into primary and secondary ensures that the maximum effort is put into solving the correct problem.

There are established techniques which are applied at this juncture for determining inter-relationships through the 'ladder of abstraction' and through 'FAST' diagramming (functional analysis system technique). A judgement of worth may also be made through a system of weighting. A detailed description of these techniques is beyond the scope of this brief text and those interested in following the concept further are refered to the bibliography.

Creativity

This stage of the job plan is the most important in that it seeks to provide alternative technical solutions to the functions identified. The normal way of generating ideas is through a process of brainstorming.

Brainstorming typically involves the formation of a group of between three and eight people from several disciplines, e.g. architect, structural engineer, mechanical and electrical engineers, quantity surveyor, maintenance manager, etc. The environment for brainstorming has to be carefully controlled, in particular superior/subordinate, client/contractor combinations within the group should be avoided. The attitude of the group members has to be right i.e. positive about the activity, curious, open minded and interested in reaching the optimal solution. The members should also be confident and communicate well, both verbally and in writing.

The group should be supported by a secretary whose function is to record the ideas as they are generated. The secretary's function is vital as ideas can easily be lost and the flow disturbed if recapping becomes necessary.

Chairmanship of the group is a key role. This person is refered to as the 'Value Management Team Co-ordinator' (VMTC), who should be skilled in the value management method and techniques and also capable of managing the brainstorming. The chairman must keep discussion on track while encouraging members to contribute and observe the rules of brainstorming which are:

- all analysis of an idea must be withheld until the judgement phase and team members must feel free to contribute ideas without criticism or any other indication of rejection;
- any idea is welcome, even a wild idea can trigger a good idea from someone else;
- it is the number of ideas which is important; research has demonstrated that the proportion of good ideas to wild ideas tends to stay reasonably constant and therefore the more ideas there are the more good ideas will be generated;
- members should be encouraged to build on others' ideas, turning them into better ideas;
- no-one should be allowed to consider their solution as the ultimate and thereby terminate or disrupt the session by continually referring to it.

It is the responsibility of the VMTC to guide the group towards the solution of the primary function. So often in design considerable ingenuity is used to solve the secondary function without consideration of the primary function. For example in the window problem described above it would be foolish to solve the problem of solar heat gain and glare without first investigating solutions to the primary functions of, prevent access, retain heat, insulate from external noise, and permit view.

At the end of the brainstorming session the secretary will have recorded many ideas which are then the subject of the judgement phase.

Judgement

During this phase the ideas generated in brainstorming are reviewed in the context of the particular building project. Ideas which are impractical are deleted as are ideas which do not accord with the client's requirements. Ideas which do not stand on their own are combined wherever possible with other ideas. Those ideas which would require technological development are set aside for reference to manufacturers.

The ideas which remain are those which are considered workable. The next exercise is to rank the ideas in order of merit which can be done in two ways. The most straightforward is carried out in the following manner.

- First, the advantages and disadvantages of each idea are determined by the team and recorded alongside the idea.
- Second, the team award 'marks out of ten' for each idea based upon a subjective judgement of the worth of the idea. The marks are summed and the ideas ranked accordingly.

The rank ordering of ideas assists in choosing those ideas which are to go forward for further development.

Alternatively, ideas may be ranked based upon the extent to which an idea meets pre-determined criteria. For example, in the deep plan office case used above, the following attributes may be considered the most important:

Initial cost
Maintenance
Energy usage
Aesthetics
Performance
Security

These attributes are then ordered according to the client's requirements. In this case the client may have indicated that initial cost is the most important aspect and that security also has been stressed. After discussion the list may be re-arranged as follows with a weighting given to each attribute:

Initial cost	(IC)	10
Security	(S)	8
Maintenance	(M)	5
Aesthetics	(A)	4
Energy usage	(EU)	3
Performance	(P)	2

The next step is to construct a chart which, on a matrix, allows the ideas to be judged on the extent to which they have the above attributes. This judgement is based on:

Poor	1
Fair	2
Good	3
Very good	4
Excellent	5

Using the example of the window above, if an alternative idea was say, an external TV camera with large flat screen monitors internally, then the chart would be as shown in Table 3.1. The chart gives a guide to those workable ideas which are to be selected for further analysis in the development stage.

Table 3.1 Judgement chart

Attributes	IC	S	M	A	EU	P
Weighting	10	8	5	4	3	2
Traditional window	good	poor	fair	very good	poor	good
	3	1	2	4	1	3
Total 73	30	8	10	16	3	6
TV Camera and Screen	fair	excellent	good	poor	excellent	good
	2	5	3	1	5	3
Total 100	20	40	15	4	15	6

Development

During development the selected ideas are taken and analysed in order to determine their technical feasibility and economic viability. This activity can be carried out by individual members of the team taking away ideas and working them up into technically viable solutions which can be costed. In addition to utilising traditional estimating techniques and skills, the quantity surveyor may also be required to prepare budget life cycle cost estimates for the developed solutions, investment appraisals, risk analysis studies, etc.

Recommendation

At this stage the team meets to discuss the technical and financial implications of each worked up idea. Those ideas which are not technically feasible or economically viable are rejected. Those which have been found to be technically sound and offer cost advantages are entered on a list of recommendations to be made to the project design team and/or client.

THE FORMAL APPROACHES TO VALUE MANAGEMENT

The following are held to be typical of the formal approaches to value management. Figure 3.1 illustrates, by reference to the RIBA plan of work, the stages at which each technique is most likely to be used.

The Charette

This method, named after the Canadian value engineer, Bob Charette, seeks to rationalise the client's brief through the identification of the function of the spaces specified.

This exercise takes place at a meeting, chaired by the value manager, involving the client's staff and the design team and aims to ensure that the latter understand fully the requirements of the former.

During research into this subject interviews were held with a client who stated that a value management study at the briefing stage is useful:

'... just to focus on what are the objectives, what are the primary functions of any given activity and this is one area of value engineering where it was clear when we got around the table that what each of them [client's staff and design team] thought was the primary function was different ...'

There is a theory which states that the brief given by the client to the design team is an amalgam of the 'wish lists' of all of the parties who contribute to the brief. This is particularly so for buildings which are to house organisations comprising diverse departments such as hospitals, universities, prisons, owner-occupied offices for complex organisations and manufacturing plants. Even where a project manager is employed there is a high probability that the brief will reflect data gathered from departmental heads who will seek to maximise their requirements. In a hospital for instance two departments may each have a requirement for a laboratory and the two laboratories may be identical, but this is not likely to be realised unless a study is made of the function of each space.

The Charette is organised along traditional job plan lines with the first stage being to gather as much information as is available regarding the function of the spaces defined in the brief. All the information necessary for the exercise resides with individual members of the team. The skill of the value manager is demonstrated in teasing this information out so that all can be equally knowledgeable about the functional aspects of the project. These functions are defined along with performance criteria, e.g. temperature requirements.

The next stage in the process is to be creative in terms of arrangement, adjacency, timetabling, etc. It may be found for example that by siting two particular hospital departments together they may use the same laboratory, with consequent savings in capital and running costs (including laboratory staff).

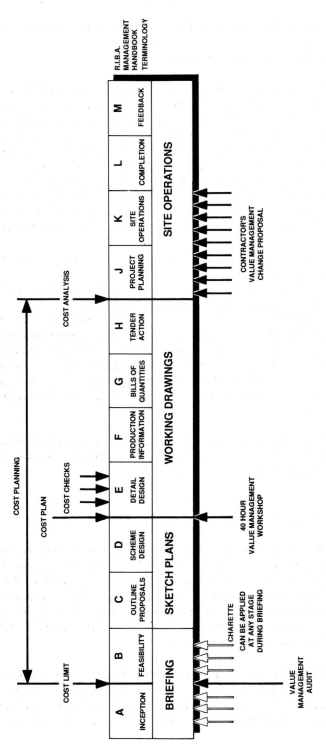

Figure 3.1 The application of value management techniques

The ideas generated during the Charette are recorded and analysed and the final decisions are incorporated into a revised brief.

The Charette has five major advantages:

- It is considered by many clients to be inexpensive. There is of course staff time to consider and the time of the design team. However, if the design team were informed at the time of their selection that this meeting was required, it is likely that it would not affect the fee bid greatly. The only significant expenditure is the fee for the value manager together with the secretary to the team who will either be provided by the client or by the value manager.
- The exercise is considered to be the best way of briefing the whole team. One industrial client with an extensive building portfolio has stated that even if the exercise did not realise any great rationalisation the very fact that all members of the team were present would mean that all understood fully his requirements.
- The exercise occurs early in the design process, stated by many to be the most cost consequent stage.
- The exercise can be carried out in a short period of time, only the most complex projects requiring more than two days work.
- Finally, the exercise cuts across organisational, political and professional boundaries. One central government organisation client has stated that a meeting of this kind would not normally be possible since departmental heads would be reluctant to give up the time, the meeting would be politically structured and the design team themselves would not normally organise such a meeting. The fact that it was a value management exercise under independent chairmanship made it happen.

The Charette is an inexpensive means of examining the client's requirements by the use of functional analysis, from which occurs facilities rationalisation together with full design team briefing.

The 40 hour value management workshop

This is the most widely accepted formal approach to value management (see Figure 3.2). Indeed the initial training of value managers as laid down by the Society of American Value Engineers (SAVE) is based on a modified 40 hour workshop. The workshop involves the review of the sketch design of a project by a second design team under the chairmanship of a value manager. It utilises all stages of the job plan within a working week and is considered to be quick and effective. The stages of the workshop are:

- The client should inform the members of the design team at the time of their fee bid that the project will be the subject of a value management exercise. This is important both from a professional relations aspect and also from the point of view of establishing how the design team are to cover the cost of any

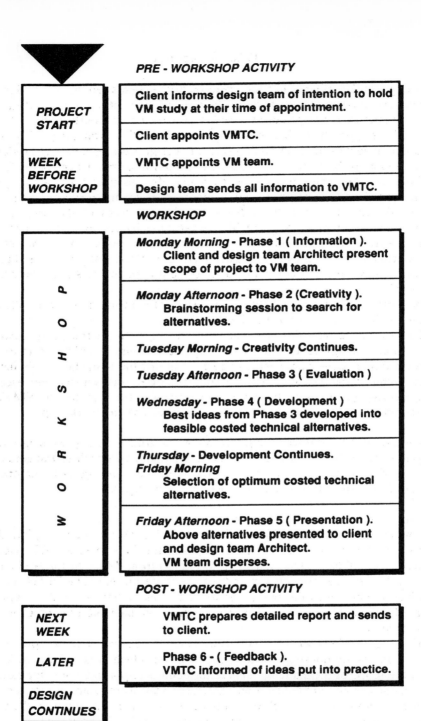

PRE - WORKSHOP ACTIVITY

PROJECT START	Client informs design team of intention to hold VM study at their time of appointment.
	Client appoints VMTC.
WEEK BEFORE WORKSHOP	VMTC appoints VM team.
	Design team sends all information to VMTC.

WORKSHOP

W O R K S H O P

Monday Morning - Phase 1 (Information).
Client and design team Architect present scope of project to VM team.

Monday Afternoon - Phase 2 (Creativity).
Brainstorming session to search for alternatives.

Tuesday Morning - Creativity Continues.

Tuesday Afternoon - Phase 3 (Evaluation)

Wednesday - Phase 4 (Development)
Best ideas from Phase 3 developed into feasible costed technical alternatives.

Thursday - Development Continues.
Friday Morning
Selection of optimum costed technical alternatives.

Friday Afternoon - Phase 5 (Presentation).
Above alternatives presented to client and design team Architect.
VM team disperses.

POST - WORKSHOP ACTIVITY

NEXT WEEK	VMTC prepares detailed report and sends to client.
LATER	Phase 6 - (Feedback). VMTC informed of ideas put into practice.
DESIGN CONTINUES	

Figure 3.2 The 40 hour workshop

redesign work arising out of the exercise. Some clients require the design team members to cover this cost within their submission, others state that they will be reimbursed for any necessary redesign work on an hourly basis.

- The client appoints the Value Management Team Co-ordinator (VMTC — the value manager) and in discussion with the design team establishes the date for the workshop. Normally the VMTC will submit a fee bid which covers the cost of the complete value management exercise described below.

- The VMTC will appoint a value management team, normally six to eight professionals, in a mix that reflects the characteristics of the project under review. For example, a project with extensive mechanical and electrical services may attract a team including four members with these professional backgrounds. Team members are drawn from professional practice and may or may not have any previous value management experience. All are paid by the VMTC.

- The workshop is normally held near the site of the proposed project, either in a hotel or in a room within the client's office.

- The date of the workshop is of key importance both for the design team and the value management team. The design team must complete their work to sketch design stage one week before the date of the workshop. This includes the architectural design and also the structural, mechanical and electrical engineering proposals. The completed drawings are sent to the VMTC for distribution to the team during the week preceding the workshop.

During the week of the workshop, the team will follow strictly the stages of the job plan detailed below. It is the logical step-by-step approach to the generation of alternative technical solutions which makes value management unique and distinguishes it from traditional cost planning techniques:

Monday morning — phase 1 (information)
Each member of the team will have had the project sketch drawings, initial cost estimate, and calculations and outline proposals for the structure and services for two days and will have gleaned some information from these. At the beginning of the workshop the design architect and the client are present. The VMTC gives an introduction and states the objectives for the week. Often the VMTC will have prepared a timetable and may also have prepared an elemental breakdown of the initial estimate.

Following the introduction the client and the design architect present the project and answer questions. The client reaffirms which areas of the project are within the scope of the exercise. This latter point is important since, for example, if the client has already reached an agreement with a trade union that a specific number of people will be employed within a plant then all ingenuity on the part of the value management team to reduce manning levels will be in vain.

After the presentation, the client and architect leave.

The team now concentrate their efforts on identifying the functions of the various parts of the building. In the study emphasis is given to those functions which attract a high cost but are not important or are secondary. Attention will also be paid to those functions which are primary and important but attract a low cost.

In one study for the modernisation of a boiler house on a large military site in North America, with an estimated project cost of $71.5 m, the team identified 17 functions of which 7 were selected for study.

Monday afternoon and Tuesday morning — phase 2 (creativity)
During this phase the group brainstorm ideas to satisfy the identified functions. In the boiler house example above over 200 ideas were generated during this session.

Tuesday afternoon — phase 3 (judgement)
At this stage the team decide which of the ideas generated are worthy of further development. For example, of the 200 or so ideas generated above only 42 were thought good enough for development. The method of selection of ideas is described earlier in this chapter.

At this point some value managers prefer to invite the design architect back to the workshop to discuss the acceptability of the ideas in principle. This can reduce abortive work if, for instance, the design team had already thought of the idea and rejected it or if the architect would not agree to such an idea under any circumstances.

Wednesday and Thursday — phase 4 (development)
During this phase the team may split as individuals or small groups to evaluate the ideas in detail. The aim is to develop the ideas into worked and costed solutions.

Friday — phase 5 (recommendation)
During the morning of the final day the group reconvenes to discuss the worked solutions. At this stage those solutions which either cost more than the original, reduce quality or are not technically feasible are rejected. In the boiler house case above 15 worked solutions were rejected at this stage leaving 27 viable solutions for presentation to the client and design architect.

In the afternoon those worked solutions accepted by the team are presented to the design architect and the client.

The formal workshop is now at an end. The members of the workshop return to their practices leaving the VMTC to take away the week's work and write the report.

The following week — action and feedback
In the early part of the following week the completed report is sent by the VMTC to the client and design architect. At this stage, for example, one North American government department takes all of the ideas, sets them out on a sheet horizontally with vertical columns for each member of the design team who receives a copy. The team members are requested to enter either 'accepted', 'rejected', or 'further discussion required' against the suggestions. A meeting is called where all members of the design team gather to discuss the suggestions. All those ideas which have been unanimously accepted are incorporated into the design. In

respect of the others, discussion takes place to determine which may be acceptable. The client will wish to be convinced of the need for rejection.

In the boiler house example above, 11 of the 27 suggestions were incorporated into the final design, leading to savings of $33m on the original estimate of $71.5m. This remarkable saving of 46% of estimated cost was achieved by demolishing two perfectly satisfactory buildings adjacent to the site and rebuilding them elsewhere. Within the original scheme the design team had used considerable ingenuity to design an expansion in the boiler house facility around the constraint of the existing buildings.

A value management workshop may therefore be considered as effective by reason of:

- The generation of alternative technical solutions to a problem which have been costed in capital and life cycle cost terms.
- The fixing of a date for the completion of a sketch design. Although not a function of the workshop it has been stated that the setting of a date for the workshop encourages the design team to complete to a more advanced stage than would otherwise be achieved.
- In the majority of cases the costs of the workshop are a small proportion of the savings achieved. Value managers comment that on any project at least 10% of the estimated contract value is within the area of unnecessary cost. They also say that the value manager will achieve a minimum 10:1 return on the investment made by the client and therefore in the majority of cases the workshop fee, usually quoted as a lump sum, would represent less than 10% of the savings realised.
- In responding to questions on 'how often do workshops fail?' value managers often state never, but one major client has stated about 2%.

The perceived problems associated with the workshop relate to conflict, time and resourcing. These centre around:

- The fact that the client may consider that the design team should arrive at the optimal solution without the need for a further exercise at additional expense. This criticism may be countered in two ways, firstly that it is the function of the design team to arrive at a workable solution given the information in the client's requirements. Secondly, that a value management study is an analysis of the ideas which have been generated. A value management study cannot be carried out until there is an idea to analyse and it is therefore truly a second phase of the design exercise. Currently, designers are not expected to carry out, nor are paid for such an exercise.
- The interpretation of the exercise by the design team as a critique of their design judgement. This is a difficult problem which is hard to counter unless the original designer plays a part in the activity. The reason given by some value managers for not including members of the original design team is the danger that established ideas are defended and their presence may stifle frank comments on the design. This potential area of conflict can be alleviated

through the education of the designers in value management techniques. Informing the members of the design team, at the time of their appointment, that the exercise is to take place and that they will be paid an additional fee for implementing design changes will also reduce conflict.

- The time frame of the value management workshop. It is beyond dispute that the value management workshop will effectively take three to four weeks from the design programme. That is one week prior to the workshop for the distribution of drawings and information, the workshop week and the period of time following the workshop for the submission of the report, discussion and design changes. In some projects this period, during which the design will be at a standstill, will be unacceptable. However, in the majority of cases it is capable of being accommodated particularly in view of the fact that the workshop itself is a watershed between sketch design and working drawings and provides an immoveable date for the completion of the sketch design.
- The resourcing of a workshop can pose problems associated with the withdrawal of professionals from their practices for a one week period. It is a condition precedent to a successful workshop that members of the value management team are isolated from their home environment for the duration of the workshop.

The workshop therefore is not without its disadvantages but has consistently proved to be a very effective means for the application of value management.

The value management audit

The value management audit is a service offered by value managers to large corporate companies or government departments to review expenditure proposals put forward by subsidiary companies or regional authorities. The procedures employed follow exactly those of the job plan. Following such a proposal the value manager visits the subsidiary company or regional authority and undertakes a study of the proposal from the point of view of providing the primary functions. The study may be carried out using personnel from the subsidiary body or others seconded from similar organisations. The study is a global review and normally takes one or two days, following which the value manager will submit a report detailing the primary objective and the most cost effective method for its realisation.

The projects director of one subsidiary company has stated that a value management audit on one proposal revealed a number of shortcomings with the statement of requirements to the extent that the company now adopts a policy of holding a Charette before a proposal is submitted to the parent company.

The contractor's change proposal

The contractor's change proposal (or value management change proposal) is a post tender change inspired by the contractor. The United States Government, for example, includes a clause in its conditions of contract which states that contractors are encouraged to submit ideas for reducing costs. If the change is accepted by the design team then the contractor shares in the saving at the rate of 55 % of the saving for fixed price contracts and 25 % for cost reimbursement contracts. For example, if a contractor, on a fixed price contract of $250 000 makes a suggestion which the contractor estimates will save $10 000, then following verification and acceptance by the design team, the contractor will receive $5500. The payment is made by reducing the contract sum by $4500.

The contract may however, be delayed while the design team investigate the viability of the change. For this reason changes tend to be relatively superficial.

VARIATIONS ON THE FORMAL APPROACHES TO VALUE MANAGEMENT

Although the four approaches detailed above are the most commonly used they are not suitable in every case. The following are applications of the job plan which have been used in practice but do not fall within the standard approaches.

The shortened workshop

In many cases the estimated project value is lower than the $2−3 million considered to be the lower limit for a full 40 hour workshop involving a team of six. In this case the 10 % rule is used to determine how much can be spent on the workshop. For example, for a project of $500 000 the target savings are $50 000 and the fee for the workshop $5000. It is now a question of determining how much professional time can be bought for $5000. If a rate of $800 per person per day is assumed, including expenses, then six person days can be afforded. This could amount to three people for two days.

The concurrent study

This approach involves the design team themselves coming together on a regular basis under the chairmanship of the value manager to review design decisions taken. The method has much to commend it, in that it answers much of the criticism levelled at the standard 40 hour workshop. The extent of involvement of the value manager needs to be determined in advance so that an appropriate fee can be established. The fee bid from members of the design team will also have to account for their extra involvement.

The concurrent study is suitable for construction management contracts in which the design is carried out in stages along with the construction (fast-tracking).

In a study of a $100m office project in Canada, a value management design meeting was held on site each Wednesday. The reference point for the comparison of costs was the elemental cost plan. At initial meetings the function of the spaces were analysed and five adjacency diagrams generated. These were reduced to three for presentation to the client.

For the selected plan a number of structural solutions were generated and analysed on a matrix along with solutions for the electrical and mechanical installations. Once the building form was established construction work began. Meetings continued with the construction manager in attendance through to the end of the project. The final cost of the project was $9m less than the budget.

A disadvantage of the concurrent study is the difficulty of proving the value of the exercise and of the time expended. Critics find it is easy to say that the budget was set too high and this is difficult to counter. The concurrent study should not be dismissed however, since if it is properly carried out under the chairmanship of a value manager then a better understanding of function and improved design can result.

DESIGN LIABILITY

The question of the design liability of the value manager hinges around whether, during the process of developing alternative technical solutions to a problem, the value manager incurs a liability for the suggestion. There are two distinct but opposing theories.

The first states that the value manager creates ideas and presents them to the design team. Whether or not the idea is incorporated into the design is a matter for the designers of the project and the client. Therefore the value manager is not liable for the idea but is liable for the professional conduct of the value management exercise.

The second theory states that if a professional puts forward a suggestion which is backed up by calculations and costings then that professional is liable for the integrity of that suggestion. If the professional is employed by the value manager then the client's recourse in the event that the suggestion fails, is to the value manager.

Each theory is valid and each theory admits some liability. The exact extent of that liability will not be established, however, until it is aired in litigation. It is interesting to note that in a survey of North American value managers none had any knowledge of such a case against a value manager.

SKILLS OF THE VALUE MANAGER

The primary skills of the value manager are those of leadership, diplomacy, chairmanship and the appropriate application of the job plan. Value managers practising in North America tend to be self-selected from a background of architecture or engineering. A construction background is important and therefore value management is seen as a post construction qualification profession.

From observation of the tasks performed by value managers an understanding of the briefing process, spatial and functional relationships, managing brainstorming, modelling of the initial estimate to permit rapid reassessment of design changes, life cycle cost modelling, agenda planning and report writing, are all of paramount importance.

Within the functional analysis area, an understanding of techniques such as the functional analysis system technique (FAST), the appreciation of worth expressed in monetary terms, and methods for making logical judgement on the value of an idea are also necessary.

Within the UK construction industry therefore, the acquisition of value management skills is most likely to be undertaken at postgraduate level following a professional qualification.

CERTIFICATION FOR VALUE MANAGEMENT

In North America the Society of American Value Engineers run a strict certification programme such that those who are Certified Value Specialists (CVS), have completed a course of examined study and run workshops on a regular basis. This certification means that those clients who employ a CVS can be confident that they are paying for the services of a trained specialist.

No such certification programme exists in the UK at present. The view has been expressed that the professional institutions should monitor the activities of their members and dissuade the advertising of services such as value management, unless those members have themselves received training or have trained personnel within their practice/company.

However the certification problem is dealt with, it is necessary that in order to inspire confident acceptance of value management by clients, those who practice it must be properly trained.

VALUE MANAGEMENT'S RELATIONSHIP WITH PROJECT MANAGEMENT.

Value management is complementary to project management. It appears likely that in the future, a project manager may wish to offer a client a value management service as a part of the whole process. Whether or not the project manager conducts the value management process is open to debate. Certainly within the professional team structure the project manager would be in a better position to conduct the exercise than anyone else.

A good professional project manager, whether employed as part of the client organisation or not, should be able to organise and carry out a successful Charette or concurrent workshop. The same is unlikely to be true of a 40 hour workshop because of the need to employ a 'second design team'.

CASE STUDY

The case study chosen to illustrate the practical application of value management was undertaken by a North American value engineer for the US Air Force on a base in Italy. The project was for the construction of a concrete parking apron adjacent to the existing runway. The parking apron was designed to be 144 × 351 m with one edge abutting the runway. The apron was designed in accordance with standard USAF requirements and the cost was estimated at $4.4m.

This study has been chosen to illustrate that even with a very simple project important factors can be overlooked and unnecessary costs incorporated.

The value management exercise was undertaken in Italy at the sketch design stage. The value engineer chose a team comprising himself as the certified value specialist, an Italian civil engineer with experience of airport construction, an Italian pavement engineer then working for a civil airport authority and a translator.

The exercise began on a Monday with an introduction of the drawings and specification. There was no presentation by the client nor by a design team representative.

During the information gathering phase it was realised that Italian aviation regulations forbid the parking of aircraft within a certain distance of the runway. This meant that 20% of the parking area as designed could only be used for access to the runway and not for parking. Since the prime function of the apron was for parking this highlighted unnecessary cost.

On Tuesday the brainstorming session took place at which 21 suggestions were made. These were summarised in the form illustrated in Table 3.2

Table 3.2. Summary of suggestions

Description	Verb:Noun	Kind Basic B Secondary S	Cost $/m²	Worth	Comments
Concrete	Support aircraft	B	44.98	35.98	20% provides clearance
Design joints	Prevent cracking	S	7.72	5.51	Spacing too close

On Wednesday and Thursday these ideas were further developed into workable solutions and presented in the form as shown in Table 3.3.

Table 3.3. Workable solutions

	Advantages	Disadvantages	Rating
Change geometric shape	Better usability	None	10
Respace joints 6 × 6 instead of 4.5 × 4.5 m	Reduction of costs	None	9
Raise apron in centre	Reduce cost of drainage	Increased risk of slip in storms	10

On Friday the result of the exercise was presented to the client. The major recommendations were:

Change geometric shape   saved $474 200
Respace construction joints   saved $110 000
Raise apron   saved $ 74 200

Although the case study is very simple it demonstrates a number of points:

- It is not necessary to have a multi-million pound project before a value management study is worthwhile.
- The original solution was perfectly workable as it was designed.
- A value management study can bring to light factors which had not previously been considered such as the Italian requirement for a safety margin alongside the runway. In this case it was better that only the access to the parking area be in concrete, the safety margin was best left as grass. The client benefited here from both better usage of the area and a less expensive solution.
- A consequence of the reduced cost of drainage was a small increase in the risk of aircraft slippage in heavy storms. The client was willing to accept this risk, particularly as such storms in the area of the base were almost unknown.

VALUE MANAGEMENT AND QUANTITY SURVEYING PRACTICE

The RICS QS Division's publication entitled *A Study of Quantity Surveying Practice and Client Demand* (March 1984) provides an indication of the extent to which quantity surveyors may offer value management services in the future. The report indicates that approximately 80 % of practices will offer only those traditional services such as the preparation of tender documentation, tender evaluation and post contract services. The remaining 20 %, often refered to as the 'leading edge of the profession', are likely to consider value management as a means of extending the services they can provide to an increasingly sophisticated construction market.

As indicated earlier in this chapter, certain skills practised by quantity surveyors, such as cost planning and life cycle costing, have an application in the provision of value management services. In addition the quantity surveyors' inherent ability to assimilate and manage data in a logical manner and ultimately to communicate such data clearly and concisely means that they are well suited to the role of value management team co-ordinator. Additionally, they can contribute to a value management study as the 'cost consultant' member of the team.

If the quantity surveyor is to play a significant role in the development of value management services in the UK construction industry it is evident that certain traditional views of quantity surveying practice must change. For example, the quantity surveyor will need to become more proficient in the practical application of cost modelling and life cycle costing techniques and also in the appreciation of the implications of proposed value management changes on the design and construction of the project. Quantity surveyors will also need to further enhance their skills in leadership, diplomacy and 'man management'. In addition the organisational structure of practice will need to accommodate the secondment of individuals to value management teams.

Within an increasingly sophisticated UK construction industry value management has an immediate potential across the entire spectrum of the industry. This is particularly true in the areas of design, construction, maintenance/facilities management and energy conservation. Current feedback from the profession indicates that some attempts are being made by quantity surveyors in private practice, contracting organisations and developer/client companies to undertake exercises in these areas under the title of value management.

CONCLUSION

The benefits of a value management service provided by chartered quantity surveyors have been demonstrated. It is also evident that other non-construction professionals such as Accountants and Management Consultants have also recognised these benefits to their construction industry clients and are currently considering the application of their skills in the context of the value management team.

Value management represents a natural progression for the quantity surveyor in leading the search for alternative feasible technical solutions and presenting them as evaluated and costed options. This can only be done however, within a structured framework of skills and techniques, used in a manner which demonstrates the provision of a totally professional service.

In value management quantity surveyors have an opportunity to develop their 'leading edge' skills and promote the profession by providing a new and innovative service to the discerning construction industry client.

BIBLIOGRAPHY

Dell'Isola, A. J. (1982) *Value Engineering in the Construction Industry*. New York: Van Nostrand Reinhold Company Inc.

Kelly, J. R. and Male, S. P. (1988) *A Study of Value Management and Quantity Surveying Practice*. London: Surveyors Publications, RICS.

Macedo, M. C., Dobrow, P. V. and O'Rouke, J. J. (1978) *Value Management for Construction*. New York: Wiley Interscience.

O'Brien, J. J. (1976) *Value Analysis in Design and Construction*. New York: McGraw-Hill.

Snodgrass, T. J. and Kasi, M. (1986) *Function Analysis — The Stepping Stones to Good Value*. University of Wisconsin.

Zimmerman, L. W. and Hart, G. D. (1982) *Value Engineering: A Practical Approach for Owners, Designers and Contractors*. New York: Van Nostrand Reinhold.

Chapter 4

Developments in Contract Price Forecasting and Bidding Techniques

MARTIN SKITMORE, *Department of Civil Engineering, University of Salford* and BRENDAN PATCHELL, *Bucknall Austin plc*

INTRODUCTION AND DESCRIPTION OF TECHNIQUES

The objective of this chapter is to describe and demonstrate the use of four types of design estimating systems that have been developed in recent years to the point of commercial use. These consist of:

- standard regression approach to item identification and pricing;
- CPS simulation for evaluation of item interdependencies;
- ELSIE, the expert system for 'front end' item and quantity generation; and
- a prototype bidding model debiaser developed by the author for 'back end' estimate adjustment for bidder characteristics.

Before considering these systems however, some preliminary observations are provided concerning the type and nature of some currently available estimating techniques.

Designers' and contractors' approaches

The functional separation of design and construction has been reflected in the development of contract price forecasting techniques largely to meet the perceived needs of both these sectors of the construction industry. For designers, the need is to inform on the expenditure implications (to the client) of design decisions to help in achieving financial targets, and value for money.

For contractors, the need is to inform on the income implications (to the contractor) should the contract be acquired. Although the nature of the competitive tendering process is such that both designers and contractors are essentially concerned with the same task — estimating the market price of the contract — the potential for contractors to access production cost information is a determining factor in the type and reliability of technique used.

Many of the methods used in practice contain a mixture of both market and production related approaches. It is common, for instance for quantity surveyors to build up rates for special items involving new materials or components from 'first principles' by finding out the manufacturer's price and making an allowance for installation and other associated costs involved in a similar manner to the contractors' estimator. It is also common for contractors' estimators to use some 'black book' method which is more intended to generate a competitive price than a genuine estimate of cost. The validity of combining the two approaches, although a popular debating point, is by no means established in academic circles. Clearly, the relationship between market prices and production costs is the determining factor, and research on this aspect has been inconclusive as yet.

It is generally accepted however that the contractors' resource based approach produces very much more reliable estimates than the designers' equivalent. Unfortunately, theoretical research in this area has until relatively recently concentrated entirely on the formulation and solution of a highly simplified mathematical construction of the contractors' bidding decision. Despite the apparent remoteness of bidding research from the real world of quantity surveying, there is an important connection with design estimating as, unlike construction estimating, bidding is directly concerned with market prices.

Strangely, bidding research has been quite opposite in many fundamental ways to estimating research. Firstly, bidding research is firmly founded in economic theory with very little empirical support whilst estimating research has no formal theoretical base at all! As a result, both research fields have run into serious logistical problems. Bidding researchers are now faced with an unmanageable number of theoretical extensions, most of which are not appropriate to real world data, and estimating researchers are now faced with an unmanageable amount of data with no theoretical basis for analysis. Secondly, bidding models are invariably probabilistic in nature whilst estimating has traditionally been treated as a deterministic matter. Perhaps the biggest emphasis in estimating research today is in the reliability of the techniques used, and statistical probability offers the greatest potential for modelling reliability. Clearly then the theories and techniques used in bidding have some relevance in estimating especially in the provision of a mathematical basis for the analysis and developments in the field.

Mathematical and typological features

Traditional estimating models can be generally represented in the form of

$$P = p_1 + p_2 + \ldots + p_N = q_1 r_1 + q_2 r_2 + \ldots + q_N r_N \tag{4.1}$$

or, more succinctly

$$P = \sum_{i=1}^{N} p_i = \sum_{i=1}^{N} q_i r_i \tag{4.2}$$

Where P is the total estimated price, p is the individual price, q the quantity of work and r the rate or value multiplier for items 1, 2, ...,n respectively. So for a bill of quantities based estimate, q_1 q_2 etc. would represent the quantities for items 1, 2 etc. in the bill, r_1 r_2 etc. the rates attached to the respective items, and p_1 p_2 etc. their individual products. For example, an item of say concrete in floors may have a quantity q of $10\,m^3$ and rate r of £60 per m^3, giving an item price p of $qr=£600$, the sum of all such prices Σp giving the total estimated price P.

The differences that occur between traditional estimating methods are usually in the number and type of items and the derivation and degree of detail involved in estimating their respective q and r values.

From Table 4.1 it can be seen that the number of items (N) may range from one single item (unit, functional unit, exponent, interpolation, floor area, cube, and storey enclosure) to very many (BQ pricing, norms, resource).

The unit method is used on all types of contracts. It involves any comparable unit such as tonnes of steelwork or metres of pipeline.

The functional unit method being similar but restricted to use on buildings with units such as number of beds or number of pupils.

The exponent method, used on process plant contracts, involves taking the contract price of an existing contract and multiplying by the ratio f some relevant quantity, such as plant capacity, of a new or existing contract, raised to a power, r, determined by previous analysis.

The interpolation method takes the price per square metre of gross floor area of two similar contracts, one more expensive and one cheaper, and interpolates between them for a new contract.

The floor area method is less specific in that a price per square metre floor area is derived somehow, depending on any information available.

Similarly the cube method involves the calculation of a quantity representing the cubic content of a building for the application of a price per cubic metre.

Perhaps the most sophisticated single item method is the storey enclosure method, involving the aggregation of major building features such as floor, wall, basements and roof areas into a composite index known as a 'storey enclosure unit'.

BQ pricing and norms methods on the other hand involve the detailed quantification of a great number of items, usually prescribed by an external contract controlling institution such as the Joint Contracts Tribunal (in the UK) or Government (in Eastern Block countries).

Several methods fall between these two extremes using either a few items (graphical, parametric, factor, and comparative) or a reduced set of BQ pricing/ norms type items (approximate quantities, elemental).

The graphical method, used on process plant contracts, involves plotting the quantities of each of a few items of interest against the contract sum of previous contracts, for a visual analysis of possibly non-linear relationships.

Parametric methods, also used on process plant contracts, adopt a multivariate approach in using a function of several process related items such as capacity, temperature, pressure in combination.

Table 4.1. Resume of estimating techniques

#	Estimate Technique	Model	Det/Prob	Relevant Contract Type	General Accuracy (cv)	Det/Prob	Items Number	Items Type	Quantities Derivation	Quantities Det/Prob	Rates Derivation Data Base	Weighting	Current	Quantity trended	Det/Prob
1	UNIT	$P = qr$	det	All	25-30%	det	single	any comparable unit, eg tonne steelwork, metre pipeline.	Brief	det	averaged price-cost unit	?	direct	none	det
2	GRAPHICAL	$P = f_r(q)$	det	Process Plant	15-30%	det	few	ditto	Brief	det	trended price-cost	?	interpolated	objective	det
3	FUNCTIONAL UNIT	$P = qr$	det	Buildings	25-30%	det	single	ditto eg number of beds number of pupils	Brief	det	averaged/ rule price-cost unit	?	direct	none	det
4	PARAMETRIC	$P = f_r(q_1\ q_2\ q_3\ \ldots)$	det	Process Plant	15-30%	det	few	process parameters eg capacity, pressure, temperature, materials, cost index	Brief	det	averaged/ rule price-cost parameter	?	direct	none	det
5	EXPONENT	$P_2 = P_1\ \dfrac{q_2}{q_1}^{\,r}$	det	Process Plant	15-30%	det	single	size of plant or equipment eg capacity	Brief	det	averaged/ rule price-cost exponent	crude subjective	direct/ interpolated	objective	det
6	FACTOR	$P = \sum_j^m fact_j\ \sum_i q_i r_i$ a) $m = 1$ (Lang method) b) $m > 1,\ fact_1 \neq fact_2$ etc (Hand method) c) $fact_j = U(\alpha_j, \beta_j)$ (Chiltern method)	det	Process Plant	10-15%	det	few	any	Brief/ measure	det	averaged/ rule price-cost	?	factored	a)none b)none c)subjective	det det det
7	COMPARATIVE	$P_2 = P_1 + \sum_i (P_{2i} - P_{1i})$	det	All	25-30%	det	few	depends on differences	Brief	det	price-cost items	crude subjective	adjusted		det.
8	INTERPOLATION	$P = qr$	det	Buildings	25-30%	det	single	gross floor area	Brief	det	price/m²	crude subjective	interpolated	-	det
9	CONFERENCE	$P = f(P_1,\ P_2\ \ldots)$	det	Process Plant	?	det	any	any	Brief/ measure	det	-	crude subjective	negotiated	-	det
10	FLOOR AREA	$P = qr$	det	Buildings	20-30%	det	single	gross floor area	Brief/ measure	det	averaged price/m²	crude subjective	direct	subjective	det
11	CUBE	$P = qr$	det	Buildings	20-45% (based on 86 cases)	det	single	volume	measure	det	averaged price/m³	crude subjective	direct	none	det
12	STOREY ENCLOSURE	$P = qr$	det	Buildings	15-30% (based on 86 cases)	det	single	floor/wall area/ basement/roof	measure	det	averaged price/SE unit	crude subjective	direct	none	det
13	BQ PRICING a)(Conventional)	$P = \Sigma\ q_i r_i$	det	Construction	10-20% (5-8% for builders)	det	v many	SMM	measure	det	a)averaged BQ's	crude subjective	direct	subjective	det
	b)(B Fine)	$P = \Sigma\ q_i r_i$	prob	Buildings	15-20%						b)$r_i = U(r_{min}, r_{max})$		direct	subjective	prob
14	SIG. ITEMS	$P = \Sigma\ q_i r_i$	det	PSA Buildings	10-20%	det	medium	SMM	measure	det	averaged BQ's/rule	crude subjective	direct	objective	det
15	APPROXIMATE QUANTITIES a)(Conventional)	$P = \Sigma\ q_i r_i$	det	Construction	15-25%	det	medium/ few	SMM combined	a)measure	det	a)averaged BQ/price book	crude subjective	composited	subjective	det
	b)(Gleeds)	$P = \Sigma\ q_i r_i$	det	Buildings	15-25%				b)Brief/ measure	det	b)averaged BQ/price book	crude subjective	composited	subjective	det
	c)(Gilmore)	$P = \Sigma\ q_i r_i$	det	Buildings	15-25%				c)Brief/ measure	det	c)averaged BQ/price book	crude subjective	composited	subjective	det
	d)(Ross 1)	$P = \Sigma\ q_i r_i$	det/ prob	Buildings	25% (based on 17 cases)				d)measure	det	d)50 BQ's averaged	none	direct	none	det

Method	Formula	det/prob	%	Type	No. items	Data/resource	det/prob	Input	BQ data	adjustment	method	objective/subjective	det/prob
e) (Ross 2)	$P = \Sigma q_i r_i$	det/prob	50% (based on 17 cases)	Buildings			det	e)measure	e)50 BQ's $r_i = a + bq_i + e$ $e \sim N(o,\sigma^2)$	none	mathematically	objective	prob
f) (Ross 3)	$P = \Sigma P_i$ $(P_i = a + bq_i + e, e \sim N(0,\sigma^2))$	det/prob	30% (based on 17 cases)	Buildings			det	f)measure	f)50 BQ's	none	mathematically	objective	prob
16 ELEMENTAL	$P = \Sigma q_i r_i$	det	20-25%	Buildings	medium	BCIS/CI sfb entities	det	Brief/measure	averaged BQ's/ BCIS/m^2	crude subjective	composited/direct	subjective	det
17 CPU	$P = \Sigma q_i r_i$	det	20-25%	Buildings	medium	Similar	det	Brief/Measure	averaged BQ	crude subjective	composited	subjective	det
18 ELSIE	$P^2 = \Sigma q_i r_i$	det		Offices	medium	DBE	det	Brief	averaged BQ/ rule	none	direct	none	det
19 NORMS (schedule)	$P^2 = \Sigma q_i r_i$	det	10-20%	Buildings	v many	SMM type eg PSA schedule	det	measure	cost based rules	none	direct	none	det
20 REGRESSION	$P = a + \Sigma q_i b_i + e$ $e \sim N(0,\sigma^2)$	det/prob	15-25%	All	few	usually contract characteristics eg gross floor area, number of storeys	det	Brief	any	crude subjective	mathematically	objective	prob
21 LU QIAN	$P = \Sigma q_i r_i$	det	?	Buildings	few	usually contract characteristics eg floor area, number of storeys	det	Brief	any	mathematically	mathematically	none	det
22 RESOURCE (Activity, operational, scheduling)	$P = \Sigma q_i r_i$	det	5-8% (builders)	All	v many	resource eg man hours, materials, plant	det	production plan	average costs	crude subjective	direct/analytical	subjective/objective	det
23 PERT-COST	$P = \Sigma P_i$ where $P_i \sim N(q_i, \sigma^2_i)$	prob	N/A	All	varies	usually time resources eg man hours	prob (time)	production plan	–	–	–	–	–
24 CPS	$P = \Sigma t_i r_i + \Sigma n_i r_i$ $t_i \sim F(\mu_i, \sigma^2_i)$	prob	6.5% (based on 4 cases)	Buildings	usually few	resource eg manhours, materials, plant	prob (time)	production plan	average cost	crude subjective	direct	none	det
25 RISK ESTIMATING	$P = \Sigma q_i r_i$	prob	N/A	Construction	usually few	any	det	any	theoretical frequency distributions of cost	crude subjective	random selection $r_i \sim F(\mu_i, \sigma^2)$	none	prob
26 HOMOGENISED ESTIMATING (BCIS on line) (BICEP etc)	$P = \Sigma q_i r_i$	det	N/A	Building	any	any	det	any	average BQ	aided subjective	direct	objective	det

Notes

F ( ) some (unspecified) probability function
N ( ) normal probability function
U ( ) uniform probability function

Factor methods, again used for process plant contracts, involve pricing only a portion of the contract which is then multiplied by a factor derived from previous similar contracts. Versions of the factor method include the Lang method, which uses a single factor; the Hand method, which uses different factors for different parts of the contract; and the Chiltern method, which uses factors given in ranges.

The approximate quantities method involves the use of composite groups of BQ pricing items which have similar quantities, e.g. floors and walls, or similar physical functions, e.g. doors and windows, with rates being derived from BQ/price book databases.

The elemental method, perhaps the first of the non-traditional approaches, also uses items representing physical functions of buildings, but with quantities and rates expressed in terms of the building gross floor area.

The reliability of estimates generated by the traditional model is a function of several factors:

(1) the reliability of each quantity value, q;
(2) the reliability of each rate value, r;
(3) the number of items, n; and
(4) the collinearity of the q and r values.

This last factor is often overlooked in reliability considerations which tend to assume that q and r value errors are independent.

Traditional thinking holds that more reliable estimates can be obtained by more reliable q values (e.g. by careful measurement), more reliable r values (e.g. by use of bigger data bases) or more items, all else being equal — a proposition questioned by Fine's (1980) radical approach to BQ pricing involving the generation of random values for both q and r values. More detailed theoretical analyses also support the view that traditional thinking may be an oversimplification. Barnes (1971), for example, investigated the implication of the proposition that different r values have different degrees of reliability, specifically that the reliability of $q_i r_i$ is an increasing function of its value. By assuming a constant coefficient of variation for each item, he was able to show that a selective reduction in the number of low valued items would have a trivial effect on the overall estimate reliability. The empirical evidence in favour of Barnes' assumption is quite strong and has culminated in the significant items method now being used by PSA.

Another break with tradition has been to develop entirely new items based on a more conceptual classification of contract characteristics. Elemental estimating is perhaps the first sophisticated example of this, involving as it does the reorganisation of traditional BQ pricing items into composite groups considered to represent mutually exclusive building functions. (A development of this using cost planning units, CPU, provides an alternative.) Most types of approximate estimating methods fall into this category, the single rate methods such as floor area, functional unit, cube, and storey enclosure, or the multiple rate methods such as approximate quantities. A further and most important characteristic of all these methods is that they were all (except for Ross', 1983, alternative approximate quantities methods and the storey enclosure method) developed in the absence of any reliability measures by which to assess their value.

Research over the last 20 years has developed with different emphasis on all of the four factors influencing estimating reliability although systems development has been centred at the item level involving:

(1) the search for the best set of predictors of tender price;
(2) the homogenisation of database contracts by weighting or proximity measures;
(3) the generation of items and quantities from contract characteristics; or
(4) the quantification of overall estimate reliability from assumed item reliability.

The first of these is typified by the regression approach, involving the collection of data for any number of potential predictors (floor area, number of storeys, geographical location, etc.) and then by means of standard statistical techniques to isolate a best subset of these predictors which successfully trades off the costs of collection against the level of reliability of estimate. The second is typified by the homogenisation aids provided by the BCIS 'on line' and BICEP systems, and the fuzzy set based automatic procedure contained in the LU QIAN system. The third is typified by the Holes', Calculix and expert systems approach such as ELSIE. This is essentially a 'front end' to a conventional estimating system where items and quantities are derived from basic project information by either a known or assumed correspondence between the two. The fourth approach, typified by the use of probabilistic (statistical) models such as PERT−COST or simulation models such as risk estimating or the construction project simulator (CPS), goes beyond the standard regression approach by introducing more complicated relationships than those assumed by the standard regression method in accommodating some interdependency between and variability within r values (PERT−COST and risk estimating) and q values (CPS).

Bidding models can be represented by

$$B = Cm \qquad (4.3)$$

where B is the value of the bid to be made by a contractor, C represents the estimated production costs that would be incurred should the contract be obtained, and m is a mark-up value to be determined by the bidder. In bidding theory the major interest is in estimating a suitable value of m which will provide the best trade off between the probability of obtaining the contract and the anticipated profit should the contract be obtained. The model for C is similar to the design estimate model for P (equation (4.2)) in that it consists of the sum of a series of quantified item and rate products, say

$$C = \sum_{j=1}^{k} q_j' r_j' \qquad (4.4)$$

where q_j' and r_j' represent the quantity and rate respectively for the jth item. If the same items are used in providing both design and construction estimates, this simplifies to, in terms of the design estimate model

$$B = \sum q_i' r_i' m = \sum q_i r_i = P \qquad (4.5)$$

as both B and P are essentially estimates of the same value — the market price of the contract.

The idea that bids and profits can be modelled in a probabilistic way, i.e. by treating them as random variables, has a long tradition in bidding theory, but it is generally assumed that it is the actual rather than estimated costs that contain the random component. Like design estimates however, it is becoming increasingly popular to treat the estimated costs also as a random component in the model. Also the similarity of approaches does not end at this point as recent empirical studies strongly suggest the existence of a close relationship between C and P.

STANDARD REGRESSION

Regression, or multiple regression analysis as it is usually called, is a very powerful statistical tool that can be used as both an analytical and predictive technique in examining the contribution of potential new items to the overall estimate reliability. Perhaps the most concentrated research on the use of regression in estimating was in a series of post graduate studies at Loughborough University during the 1970s and early 1980s. The earliest of these studies developed relatively simple models with new items to estimate tender prices of concrete structures, roads, heating and ventilating installations, electrical installations and offices. More recent studies at Loughborough have been aimed at generating rates for standard bill of quantities type items by regressing the rates for similar items against the quantity values for those items.

The regression model usually takes the form

$$Y = a + b_1 X_1 + b_2 X_2 + \ldots + b_n X_n \qquad (4.6)$$

where Y is some observation that we wish to predict and X_1, X_2, $\ldots$, X_n are measures on some characteristics that may help in predicting Y. Thus Y could be the value of the lowest tender for a contract, X_1 could be the gross floor area, X_2 the number of storeys, etc. Values of $a, b_1, b_2, \ldots, b_n$ are unknown but are easily estimated by the regression technique given the availability of some relevant data and the adequacy of some fairly reasonable assumptions about the data.

The regression model in equation (4.6) is quite similar to the estimate model in equation (4.1), as Y, X, and b values can be thought of as representing the P, q, and r values. The major difference in approach however is that in applying the regression technique *no direct pre-estimates are needed of the values of item rates, r*, as these are automatically estimated by the technique. This then obviates the need for any data bank of item rates thus freeing the researcher to examine any potential predictor items for which quantities are available. The implications of this are quite far reaching for without the need to have an item rate database the research task is simply an empirical search for the set of quantifiable items which produces the most reliable estimates for Y.

Problems and limitations

Although the task appears to be straightforward, several problems have been encountered. These problems concern the model assumptions and limitations, data limitations, and reliability measures.

Model assumptions
The major assumptions of the basic regression model are that

(1) the values of predictor variables are exact;
(2) there is no correlation between the predictor variables;
(3) the actual observations Y are independent over time; and
(4) the error term is independently, randomly and identically normally distributed with a zero mean.

In terms of equation (4.1) this implies:

- that the quantities q are exact rather than approximate quantities;
- that quantities for items do not change in tandem, as does floor area and number of storeys or concrete volume and formwork area for example:
- that the tender prices for one contract are not affected by the tender prices for the previous contract; and
- that the differences between the regression predictions and the actual tenders are purely unaccountable random 'white noise', unrelated in any way to the variables used in the model.

Violation of these assumptions is not necessarily fatal to the technique. The type and degree of effects of violations depends on the type and degree of violation. Unfortunately however this is a rather specialist area in which statistical theory is not yet fully complete. The usual pragmatic approach to this is to try to minimise violations by a combination of careful selection of variables and tests on the degree of resulting violations.

Data limitations
Although there is no theoretical limit to the number of predictor variables that may be entered into the regression model, at least one previous contract is needed for every variable. For reasonably robust results it is often recommended that the number of previous contracts is at least three times the number of variables in the model. Thus for the full traditional bill of quantities model of say 2000 items this means about 6000 bills of quantities would be needed for analysis, even assuming that each bill contains identical items. Fortunately however there is a kind of diminishing return involved in the introduction of each new variable into the model so that there comes a point at which the addition of a further variable does not significantly contribute to the reliability of the model.

This property of regression analysis is often utilised by researchers in a technique called forward regression which involves starting with only one variable in the model. A second variable is then added and checked for its contribution. If it is significant, a third variable is added, checked and so on until a variable is encountered that does not significantly contribute. This variable is then left out of the model and the analysis is completed.

An extension of this method is stepwise regression which leaves out any non-significantly contributing variable already in the model and enters significantly contributing ones until completion. Although stepwise regression works fine if the predictor variables are not correlated as required by model assumption (2) above, violations of this assumption result in different models depending on the order of variables entered and removed from the model. This can be overcome by yet another technique called best subset regression, which examines all possible combinations of predictor variables for their joint contribution, selecting the best set of predictor variables which significantly contribute.

A key issue of course is in specifying a criterion for distinguishing between significant and non-significant contributions to the model. In regression formulations it is usual to concentrate on the behaviour of the error term, i.e. the difference between the actual values of Y and those predicted by the regression model. If for example ten contracts have been used in the analysis, then there will be ten actual lowest tenders (contract values) and ten model predictions. The differences between each of these ten pairs of values is then squared and added together, the resulting total being called the residual sums of squares (RSS). As each new variable is entered into the equation, the RSS decreases a little.

Two possible significance criteria therefore are the minimum total RSS or the minimum decrease in RSS as a new variable is entered. Clearly if this figure is set at zero, then all variables will be entered into the model. If on the other hand the minimum is set at some high level, very few, if any, variables will be entered into the model. Another possibility is to use the proportion of RSS to the total sums of squares that would be obtained if no variables were entered into the equation (TSS). Most standard regression packages use this latter method. For construction price−cost estimates this may not be appropriate. The decision is ultimately an arbitrary one. In most construction price−cost estimating research, the number of variables entered into the model before cut off using various sensible criteria levels is seldom more than ten, which suggests that data limitation is not likely to be a serious problem in practice.

Reliability measures

This has been perhaps the biggest area for problems and misunderstandings in all of regression based research to date. The standard test statistic given by regression packages is the F value which is a measure of the proportion of RSS to TSS mentioned above. This tests the assumption (null hypothesis) that the predictor variables used in the model have no real predictive ability, such apparent ability revealed by the regression technique being more attributable to chance than some underlying correlation. Thus if we use the F value to test a one variable model regressing contract value against gross floor area it is incomprehensible that the null hypothesis should hold, and there is no practical advantage in testing against the incomprehensible.

Another very common failing has been to confuse measures of a model's fit with measures of the model's predictive ability. Measures of the extent to which the model fits the data are readily available in most standard regression packages, the most popular measures being the F test mentioned above which offers evidence on whether the model's fit is due to chance, and the multiple correlation coefficient which indicates the degree to which the model fits the data. However, as the model has been derived from the same data by which it is being tested for fit, it is not at all surprising that the fit should often be a good one.

The real test of reliability of the regression model is to see how it performs in predicting some new data. The obvious way of doing this is to obtain the regression a and b coefficients from the analysis of one data set, collect some more X data for some more contracts, apply the old a and b values to obtain estimates of Y, and then use these estimates against the actual values of Y as a means of measuring the model's predictive ability. A more subtle approach, called jackknife validation, is to omit one contract from the data, calculate the a and b coefficients, estimate the Y value of the omitted contract and compare this with the actual Y value for that contract. This procedure is then repeated for all the contracts and then the residual analysis is conducted as before.

Applications

The regression method involves six operations:

(1) data preparation and entry;
(2) selection of model;
(3) selection of predictor variables;
(4) estimation of parameters;
(5) application of parameter estimates to specific task;
(6) reliability analysis.

In practice, operations (2) and (3) are executed concurrently.

Data preparation and entry
Data are prepared in the form of a matrix in which the rows correspond to contracts and columns correspond to the lowest tender (contract value) and values of potential predictor variables. Most regression packages can handle a few missing data which need to be flagged by the use of a special number such as 0 or 999. Contract values are normally updated by one of the tender price indices although this is not strictly necessary (the tender price index applicable to each contract could be entered as a predictor variable for instance). Another possibility is to include some approximate or even detailed estimate of the contract value as a potential predictor variable also. If the latter is used then the regression is

essentially a *debiasing* rather than estimating technique (these are examined later).

It is advisable to carefully check that the data is correctly prepared and entered before analysis as errors may not be immediately apparent. If errors do exist in the data, it is quite likely that a great deal of time and effort could be wasted in abortive analysis prior to their correction. Most regression packages offer a facility to reproduce the data in hard copy form for checking purposes.

Selection of model and predictor variables

These operations are very well documented in many intermediate level statistics texts and regression package manuals. It is usual to carry out both operations concurrently, seeking the best model and subset of predictor variables together. Two fundamentally different approaches exist that have great significance for academic work. One, called the deductive approach, involves the proposition of specific pre-analysis (a priori) hypotheses based on some theoretical position derived from an examination of extant ideas on the subject. The consequent models and predictor subsets are then tested against each other for primacy, any potential new models and subsets, however obvious from the data, being strictly excluded.

The other approach, called inductive, is strictly empirical in that the intention is to find patterns in the data that can be used to generate some future hypotheses and, hopefully, theoretical foundations. For all practical purposes the separation of these approaches is hardly relevant except as a stratagem for dealing with the logistical problems that are invariably encountered in regression analysis.

Construction price−cost estimation lacking any formal theoretical base tends to preclude the deductive approach and we are usually left to look for patterns in the data. This means trying out many possible models using not only the simple additive models described here but those involving transformations of variables (e.g. log, powers, reciprocals) together with combinations of several variables (e.g. products, powers) in either raw or transformed states. Even with only two predictor variables the number of combinations of transformations and combinations are quite substantial, with a large number of variables, some simplification is needed. In the absence of theory such simplification is bound to be an arbitrary process. As a result, research in regression is far from comprehensive and, because of the arbitrary means employed, has not been reported well enough to allow any incremental progress to be made. Also, with the confusion over reliability measures, it has not been possible to properly evaluate the progress that has been made.

Experience suggests that the best and easiest starting point is the simple regression model with raw (untransformed) values and no interaction terms (combinations of variables). Then, by using stepwise or best subset regression, this model can be trimmed down to significant predictors. The next stage is to try to find any other variable set or transformation that will significantly improve the model.

Parameter estimation

Once a satisfactory model is fitted to the data, the standard regression package will automatically calculate values of the a and b coefficients for use in estimating.

This an extremely simple process by which the value of the predictor variables for the new contract are just multiplied by the b coefficients and added together with the a coefficients. The resulting total is the regression estimate for the contract.

Reliability analysis
The reliability of the regression model can be estimated in terms of both bias, i.e. the average error of the forecast, and variability, i.e. the spread of the forecast errors. The regression technique is designed to give unbiased predictions and therefore the average error of prediction is assumed to be zero. Two relevant measures of the likely variability of the forecast are the 95% confidence limits and the coefficient of variation (see Appendix A).

Table 4.2. Regression data

Contract number	Standardised contract sum *	Gross floor area (m²)	Number of storeys	Air-conditioning (0 = no) (1 = yes)	Contract period (months)	Number of tenders
CASENO	CONSUM	GFA	STOREY	ACOND	PERIOD	BIDDERS
1	1 085.95	452	2	0	8	6
2	5 042.91	1 601	7	0	11	8
3	2 516.59	931	3	1	11	7
4	18 290.60	6 701	7	1	17	6
5	3 195.81	219	3	0	12	1
6	8 894.68	3 600	6	0	15	9
7	932.06	490	2	0	7	6
8	979.93	415	1	0	8	8
9	1 684.94	504	3	0	9	6
10	1 896.39	320	2	0	7	7
11	8 789.05	372	2	0	6	7
12	2 445.12	837	2	0	9	4
13	1 501.91	491	3	0	6	6
14	1 114.31	496	1	0	6	6
15	943.48	430	2	0	99	99
16	3 670.98	1 368	4	0	12	4
17	1 094.75	469	2	0	6	4
18	4 584.87	1 260	2	0	8	5
19	10 942.28	2 994	8	1	15	1
20	760.29	312	2	0	6	6
21	3 002.67	1 225	2	0	9	7
22	2 720.44	1 230	2	0	10	8
23	58 365.39	23 089	7	1	20	7
24	11 323.40	4 273	4	1	20	7
25	37 357.91	11 300	5	1	18	6
26	46 309.12	14 430	3	1	30	6
27	1 704.17	437	3	1	10	9
28	6 792.04	2 761	5	1	12	6

* Contract sum ÷ tender price index ÷ location factor e.g. contract number 2 contract sum = £1514385 in London (location factor 1.30) on Sept 1987 (TPI 231). Standardised contract sum = 1514385 ÷ 231 ÷ 1.30 = 5042.91

Example

Table 4.2 contains a set of data extracted from the RICS Building Cost Information Service's Brief Analysis files. A total of 28 contracts are included for office blocks from a variety of geographical locations in the UK for the period 1982 to 1988. The dependent variable contract sum (CONSUM) was standardised by dividing by the all-in tender price index and location factor to remove inflationary and location effects. The independent (predictor) variables chosen were the gross floor area (GFA) in square metres, the number of storeys (STOREY), air-conditioning (ACOND) valued at 1 if present and 0 if not present, contract period (PERIOD) in months, and the number of tenders (BIDDERS) received for the contract. The contract period and number of tenders received was not known for contract number 15 and these were given a special 'missing' value of 99.

The problem now is to provide an estimate for a new contract, not included in the data set, which has a gross floor area of 6000 square metres, 5 storeys, no air-conditioning, an 18 month contract period and 6 tenders.

The first task is to fit a suitable model to the data set. From experience and a few trials a reasonable model was found by (1) using the price per square metre value (CONSUM/GFA ratios), and (2) taking the natural logs of all the variables except ACOND.

The stepwise procedure was used to enter the independent variables into the regression model one at a time. The resulting sequence of independent variables entering the model was firstly BIDDERS, then GFA, PERIOD, STOREY, and finally ACOND. Table 4.3 gives the results obtained as each variable was entered into the model.

Up to step 5 the results are very much as expected with the number of bidders and gross floor area producing a negative regression coefficient indicating a drop in price per square metre as these variables increase due to intensity of competition and economies of scale respectively. Similarly the contract period and number of storeys produce a positive regression coefficient indicating a rise in price per square metre as these variables increase due to the effects of complexity of design and construction. The negative value for air-conditioning at step 5 however is not expected, as it indicates that air-conditioned buildings are cheaper. This, together with the increase in variability as measured by the 95% confidence limits and coefficient of variation suggests that the ACOND effect is spurious. The best looking model here seems to be that obtained at step 3, which has a better standard error of forecast and better coefficient of variation than the other models.

The model at step 3 is

$$\log (CONSUM/GFA) = a + b_1 \log (BIDDERS) + b_2 \log(GFA) + b_3 \log(PERIOD) \tag{4.7}$$

where a b_1 b_2 b_3 are the constant and regression coefficients respectively. Thus the forecast of log (CONSUM/GFA) for the new contract is, from Figure 4.1 (b):

Table 4.3 Stepwise regression results

Variables entered	Regression coefficient	Constant	Forecast	95% confidence limits for new forecast	cv
Step 1					
BIDDERS	−0.4355	1.8298	1.0495	±0.6844	38.31
Step 2					
BIDDERS	−0.4251				
GFA	−0.0395	2.0929	0.9873	±0.7090	39.10
Step 3					
BIDDERS	−0.2987				
GFA	−0.3071				
PERIOD	0.8814	1.7040	1.0609	±0.6011	32.05
Step 4					
BIDDERS	−0.2665				
GFA	−0.3293				
PERIOD	0.8480				
STOREY	0.1032	1.7887	1.0634	±0.6075	33.99
Step 5					
BIDDERS	−0.2646				
GFA	−0.3253				
PERIOD	0.8848				
STOREY	0.1108				
ACOND	−0.0667	1.6847	1.1166	±0.6836	35.06

$$\begin{aligned}
\log(\text{CONSUM/BIDDERS}) &= 1.7040 - 0.2897 \times \log(6) - 0.3071 \times \log(6000) + \\
&\quad 0.8814 \times \log(18) \\
&= 1.7040 - 0.2897(1.7918) - 0.3071(8.6995) + \\
&\quad 0.8814(2.8904) \\
&= 1.0609
\end{aligned}$$

The antilog of 1.0609 is 2.8890 which is the standardised pounds per square metre estimate, and the standardised total estimate is therefore $2.8890 \times 6000 = 17\,334$. This can now be converted into a current estimate in a given location by multiplying by the current tender price index and location factor. If the new contract is for January 1989 (TPI = 300) in Salford (Location Factor = 0.90), the estimate will be $17\,334 \times 300 \times 0.90 = £4\,680\,180$.

The 95% confidence limits of £2 565 918 and £8 536 266 (see Appendix A) are very large — a reflection to some extent on the limited amount of data used in this example.

CONSTRUCTION PROJECT SIMULATOR

The construction project simulator (CPS) was developed between 1980 and 1984 at the University of Reading by John Bennett and Richard Ormerod. It is a

***** M U L T I P L E   R E G R E S S I O N *****

Listwise Deletion of Missing Data
    Selecting only Cases for which   CASENO LE   28.00

            Mean   Std Dev Label

PRICE   1.088   .395
GFA ($\bar{x}_2$)   7.103   1.279
STOREY   1.082   .566
ACOND   .333   .480
PERIOD($\bar{x}_3$) 2.336   .436
BIDDERS($\bar{x}_1$) 1.703   .536

N of Cases =   27

Figure 4.1(a) Regression output at step 3

21 MAR 89   example of regression analysis estimating for bcis offices
09:55:07   University of Salford   Prime 9955   rev 20.1.0

* * * * M U L T I P L E   R E G R E S S I O N * * * *

Equation Number 3   Dependent Variable.. PRICE

Descriptive Statistics are printed on Page 3

Beginning Block Number 1. Method: Enter   BIDDERS   GFA   PERIOD

Variable(s) Entered on Step Number 1.. BIDDERS
                                    2.. PERIOD
                                    3.. GFA

Analysis of Variance

              DF   Sum of Squares   Mean Square
Regression   3   2.28993   .76331
Residual   23   1.75910 (s^2) .07648

F = 9.98020   Signif F = .0002

Var-Covar Matrix of Regression Coefficients (B)
Below Diagonal: Covariance   Above: Correlation

              BIDDERS   PERIOD   GFA
BIDDERS (s^2_{11}) .01211   .37795   -.38235
PERIOD (s^2_{13}) .01126 (s^2_{33}) .07333   -.88752
GFA (s^2_{12}) -.00390 (s^2_{32}) .02226 (s^2_{22}) .00858

------------- Variables in the Equation -------------

Variable   B   SE B   Beta   T   Sig T
BIDDERS (b_1) -.289E83   .110059   -.393112   -2.632   .0149
PERIOD (b_3) .881413   .270786   -.974902   3.255   .0035
GFA (b_2) -.307112   .092623   -.995027   -3.316   .0030
(Constant)(α) 1.704044 .357162   4.771   .0001

End Block Number 1   All requested variables entered.

Figure 4.1(b) Regression output at step 3

**** MULTIPLE REGRESSION ****

Equation Number 3   Dependent Variable..   PRICE

Casewise Plot of Standardized Residual

*: Selected   M: Missing   X: Unselected

Case #	−3.0 .... 0.0 .... 3.0	PRICE	*SDRESID	*DRESID	*ADJPRED	EXP ADJPRED/PRICE	EXP −DRESID
1		.88	−.9860	−.2823	1.1588		1.33
2		1.15	−.7307	−.2103	.9370		0.81
3		.99	−.5942	−.1737	1.1681		1.19
4		1.00	.1031	.0309	.9733		0.97
5		2.68	3.0287	1.1730	1.5075		0.31
6		.90	−.1312	−.0392	.9437		1.04
7		.64	−1.3530	−.3909	1.0239		1.46
8		.86	−.8505	−.2501	1.1093		1.28
9		1.21	−.0137	−4.0407E−03	1.2109		1.00
10		1.78	3.0960	.7677	1.0117		0.46
11		.75	−.5638	−.1671	.9190		1.18
12		1.07	−.3665	−.1054	1.1785		1.11
13		1.12	.9900	.2722	.8259		0.75
14		.81	−.1842	−.0555	.8650		1.06
15		.79					—
16		.99	−1.0747	−.3048	1.2919		1.36
17		.85	−.5630	−.1722	1.0198		1.19
18		1.29	1.6465	.4669	.6248		0.63
19		1.30	−1.7940	−.6667	1.9628		1.95
20		.89	−.4103	−.1216	1.0123		1.13
21		.90	.0121	3.5238E−03	.8930		1.00
22		.79	−.5588	−.1615	.9553		1.18
23		.93	−.9884	−.3222	.6052		0.72
24		.97	−.9481	−.2892	1.2637		1.34
25		1.20	1.3244	.3945	.8013		0.67
26		1.17	−.3167	−.1056	1.2716		1.11
27		1.36	−.5284	−.1692	1.1927		0.85
28		.90	−.1523	−.0444	.9446		1.05
29	X						—
		−13.30	−49.4444	−14.3656	1.0609		
Case #	−3.0 .... 0.0 .... 3.0	PRICE	*SDRESID	*DRESID	*ADJPRED		

$\bar{x}$ = 1.0419
σ_{n-1} = 0.3339
cv. = 32.05%

Figure 4.1(c) Regression output at step 3

21 MAR 89   example of regression analysis estimating for bc's offices
09:55:07   University of Salford   Prime 9955   rev 20.1.0

* * * * M U L T I P L E   R E G R E S S I O N * * * *

Equation Number 3   Dependent Variable..   PRICE

Residuals Statistics:
Selected Cases: CASENO LE   28.00

	Min	Max	Mean	Std Dev	N
*PRED	.6952	2.2392	1.0880	.2969	27
*ZPRED	-1.3234	3.8792	.0000	1.0000	27
*SEPRED	.0650	.2184	.0994	.0388	27
*ADJPRED	.6052	1.9628	1.0693	.2608	27
*RESID	-.3548	.6954	.0000	.2691	27
*ZRESID	-1.2829	2.5145	.0000	.9405	27
*SRESID	-1.7132	2.6420	.0252	1.0928	27
*DRESID	-.6667	1.1730	.0187	.3751	27
*SDRESID	-1.7940	3.2960	.0585	1.1846	27
*MAHAL	.4749	15.2555	2.8889	3.5218	27
*COOK D	.0000	2.8055	.1564	.5474	27
*LEVER	.0183	.5369	.1111	.1355	27

Total Cases =   28

Unselected Cases: CASENO GT   28.00

	Min	Max	Mean	Std Dev	N
*PRED	1.0609	1.0609	1.0609	.0000	1
*ZPRED	-.0913	-.0913	-.0913	.0000	1
*SEPRED	.0891	.0891	.0891	.0000	1
*ADJPRED	1.0609	1.0609	1.0609	.0000	1
*RESID	-14.3656	-14.3656	-14.3656	.0000	1
*ZRESID	-51.9448	-51.9448	-51.9449	.0000	1
*SRESID	-49.4444	-49.4444	-49.4444	.0000	1
*DRESID	-14.3656	-14.3656	-14.3656	.0000	1
*SDRESID	-49.4444	-49.4444	-49.4444	.0000	1
*MAHAL	1.5725	1.5725	1.5725	.0000	1
*COOK D	.6164	.6164	.6164	.0000	1
*LEVER	.0582	.0582	.0582	.0000	1

Durbin-Watson Test =   1.00664 (Sig. pos. corr.)

Figure 4.1(d) Regression output at step 3

resource type system, but with a unique facility to model some special aspects of uncertainty particularly prevalent in building work — productivity variability and external interferences to the construction process on site. Data describing a particular project — bar charts, direct and indirect costs, resources, weather, and productivity — is input and fed to a series of stochastic simulation programs which compute cost and time estimates for the project. The result is a histogram of time and associated costs which can be used as a measure of reliability of the estimates.

The system is based on the model

$$P = \sum_{i=1}^{n} t_i \, r_i + \sum_{j=1}^{m} u_j \, r_j \qquad (4.8)$$

where t is a stochastic (random) variable, $t_i \sim F(\mu_i, \sigma_i)$, representing the time taken to perform the ith activity, u represents the remaining non-stochastic quantities such as materials, and r the appropriate unit rates for converting the quantities into costs. The actual model used by the system is rather more sophisticated than this as activities may be delayed by stochastic interferences and the overall effects automatically rescheduled to enable dependent items such as preliminaries to be costed.

Problems and limitations

Model assumptions
Although the CPS is undoubtedly superior in many respects to similar resource based systems, the basic model contains two particular limitations at the present time. Firstly, in assuming stochastic independence, any actual *interdependences* between future activity times are currently incompletely modelled by the system. Thus, under the model assumptions, any time slippage caused by some future chance delay could only be recovered by some equally chance future hastening event, whilst it is conceivable that a work team or manager may take direct remedial action in the live situation. Similarly, the model makes no assumption that a chance exceptional productivity rate on one activity would be accompanied by a similarly exceptional rate on another activity, although this situation may well arise in reality due to some common characteristics of the people involved in both activities. Secondly, the non-stochastic components in the model — material quantities and unit rates — are also known to be variable to some degree and therefore may be better treated in a stochastic manner. The research and limited tests on the system to date however suggest that neither of these modelling limitations significantly affects the predictive abilities of the system.

Data limitations
As with all resource based systems, the estimation of likely resource demands requires a knowledge that is usually restricted to the constructor. Thus the very nature of design and construct, concerned with the product and the process respectively, militates against the use of the CPS by designers.

In gathering data on variability it is desirable to compare like with like. The quantity and quality of work carried out, and the method of execution should be identical or at least very similar. Bennett and Ormerod's attempts to quantify variability and differentiate between different activities however have been plagued by a dearth of relevant information. Their major source is the Building Research Establishment although even this has limitations on the type of building (traditional housing) and accuracy.

Relevant data on interferences is also severely restricted at present mainly due to the relatively recent recognition of interference as a major contributory factor. The extent of subcontractor non-attendance or labour supply problems for instance has not been accurately recorded by contractors, although some information on this is, somewhat surprisingly, available from designer sources and the management of interferences literature.

Perhaps the best and most reliable source of data is that issued by the Meteorological Office on the seasonal frequency of weather conditions. These can be obtained in the form of frequency distributions, an ideal representation for stochastic simulation.

Reliability measures
This is certainly the strongest feature of the CPS system. The system's facilities include an excellent graphical display of both histogramic and cumulative frequencies of time and costs, enabling the probability of a particular cost being exceeded to be obtained at a glance.

Applications

The system has been developed specifically for ease of use by non-experts with a limited amount of computer equipment at their disposal. The main applications features concern the entry of appropriate bar charts, unit costs, resourcing and frequency distributions, and the simulation procedure.

Bar charts
The CPS bar charts are arranged as a hierarchy consisting of a primary level in which the user defines separate activities or primary work packages (PWP) such as substructure, external envelope, internal subdivisions, etc., and a secondary level or secondary work packages (SWP) defining constituent activities of PWPs such as excavate over site, excavate bases, mass fill bases, etc.

The bar charts are created directly on the computer screen by modifying charts for previous projects — mainly the start and end dates — up to a maximum of 39 bars each for PWPs and SWPs.

The logical links (maximum 250) acting as constraints to progress are next entered by the user and drawn on the screen as thin lines. Holiday periods (maximum 9) can also be entered in much the same manner as the bars are entered. Following entry or correction, the bars and links are automatically rescheduled before any further processing commences.

Unit costs

On completion of the bar charting, the user is presented with a tabulated version of the activities against which unit costs are displayed for labour and materials separately. Temporary work activities (preliminaries) and their associated costs are entered separately by a special program facility.

Resources

Resources are entered in a way common to most resource estimating packages by means of a resource library or database. Typical trades and unit costs are contained in 100 categories. Gang compositions are built up in a separate file and these are then attached to the activities contained in the SWPs with the aid of a further program for the purpose. Various displays are available to enable the user to experiment with different combinations and quantities of resources to obtain the best mix.

Frequency distributions

A wide variety of theoretical distributions (uniform, triangular, normal, beta, etc.) and empirical distributions derived from live data are available through one of the CPS databases. The user selects one of these distributions and enters it against each activity together with the appropriate parameters for the theoretical distributions. Interference factors are dealt with in a similar manner by entering a number between 0 and 99 to define the percentage chance that an interference

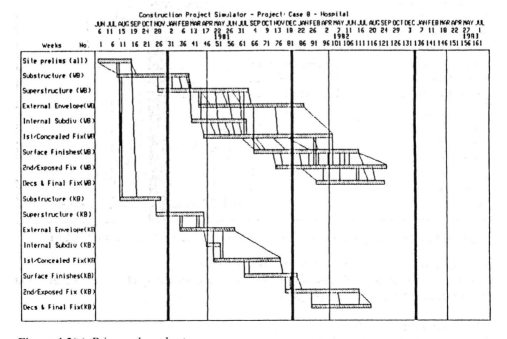

Figure 4.2(a) Primary bar chart

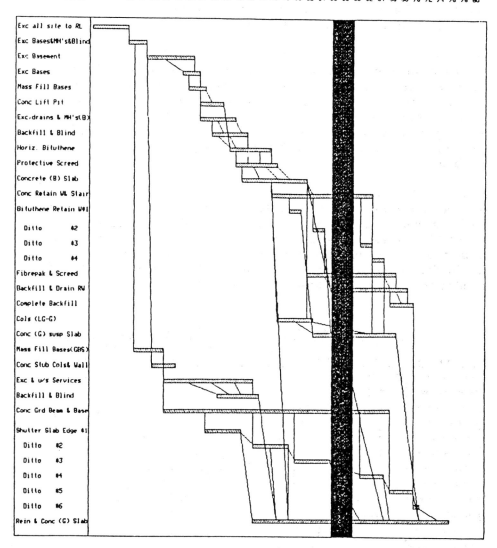

Figure 4.2(b) Secondary level bar chart (extended plan of PWP 2 in Figure 4.2(a))

will occur at any link position on an activity bar.

Twelve location specific histograms for the weather for each month of the year are entered into a separate program.

The simulation procedure

Simulation can be carried out at various stages, so that data input can have

```
PWP No. 2              AMEND COST DATA        Substructure (WB)
Activity               Labour Cost   Material Cost
Exc all site to RL     <11610  >     <9395   >
Exc Bases&MH's&Blind   <8645   >     <6680   >
Exc Basement           <14925  >     <12875  >
Exc Bases              <6635   >     <5365   >
Mass Fill Bases        <520    >     <1300   >
Conc Lift Pit          <1615   >     <3565   >
Exc,drains & MH's(B)   <7900   >     <11705  >  LABOUR TOTAL
Backfill & Blind       <1065   >     <450    >    <      118520>
Horiz. Bituthene       <775    >     <1590   >
Protective Screed      <440    >     <905    >  MATERIAL TOTAL
Concrete (B) Slab      <3500   >     <7735   >    <      160300>
Conc Retain W& Stair   <4040   >     <8920   >
Bituthene Retain W#1   <250    >     <455    >  GRAND TOTAL
    Ditto      #2      <245    >     <455    >    <      278820>
    Ditto      #3      <245    >     <455    >
    Ditto      #4      <245    >     <455    >
Fibrepak & Screed      <1445   >     <2955   >
Backfill & Drain RW    <12020  >     <10725  >
Complete Backfill      <3510   >     <2845   >
Cols (LG-G)            <2150   >     <4755   >
```

Figure 4.2(c) Cost screen

different degrees of detail and simulations still performed. A simulation of the primary level only may be carried out after the primary bar chart and weather data, and optionally the preliminary schedule and primary cost table, have been entered. This allows a quick assessment of schemes at an early stage in their development, or at tendering stage.

A simulation of one or more secondary plans may be carried out after a secondary bar chart, and optionally the secondary cost table and resources details have been entered. This allows the assessment of single PWPs when design details become available as the scheme develops, or allows the reassessment of a PWP after some change to the construction method.

A full scale simulation of all the secondary plans, followed by a simulation of the primary level can be carried out automatically. This allows the assessment of the whole scheme, at a level consistent with a good contract programme.

The manner in which this is achieved is via the use of pseudo random numbers (RNs) generated by the computer to the frequency distributions specified. The procedure used by the program on one simulation iteration is to take each bar segment defined by logical links in turn and to choose an actual duration via RNs, the assigned distribution and the variability amount. Each link position is then examined to see if an interference occurs there and the duration of the interference is chosen by an RN if an assigned interference distribution is present. This process is repeated for each bar and the whole chart is then rescheduled by the logical restraints. The cost of each bar is then calculated from the ratio of the actual bar duration to the original bar duration multiplied by the labour cost and added to

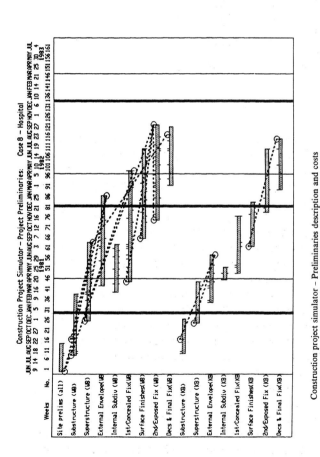

Construction project simulator – Preliminaries description and costs

	Week start		Week end	Description	Labour	Materials
1. Site prelims (all)	(0)			Staff, hutting, power, clean, other, profit	2480	333 285
			(115) to 2nd/Exposed Fix (WB)			
2. Site prelims (all)	(0)		(0) to External Envelope (WB)		380	0
3. Substructure (WB)	(16)		(81) to Superstructure (WB)	Structural foreman 2 No.	655	5150
4. Substructure (WB)	(8)		(60) to 1st/Concealed Fix (WB)	Tower Crane No. 1	315	0
5. Substructure (WB)	(8)		(92) to Superstructure (WB)	Assistant engineer + chain boy	400	0
6. Superstructure (WB)	(60)		(60) to Decs & Final Fix (WB)	Engineer + assistant	255	800
7. Superstructure (WB)	(25)		(110) to External Envelope (WB)	Hoists 2 No.	890	0
8. 1st/Concealed Fix (WB)	(42)		(81) to 2nd/Exposed Fix (WB)	Scaffold	230	0
9. Surface Finishes (WB)	(62)		(115) to 2nd/Exposed Fix (WB)	Services coordinator	485	0
10. 2nd/Exposed Fix (WB)	(71)		(115) to 2nd/Exposed Fix (WB)	Finishing foremen 3 No.	205	0
11. Substructure (KB)	(16)		(54) to External Envelope (KB)	Completion agent	660	4650
12. Superstructure (KB)	(24)		(54) to External Envelope (KB)	Tower Crane No. 2	155	0
13. Surface Finishes (KB)	(58)		(108) to Decs & Final Fix (KB)	Scaffold	165	0
				Finishing foreman		

Figure 4.2(d) Preliminaries schedule and cost table

CREATE RESOURCE SETS

TITLE	COST
23. ⟨Grdwk bricklyr ⟩	⟨150 ⟩
24. ⟨Grdwrk plantdvr⟩	⟨120 ⟩
25. ⟨Drainlayer ⟩	⟨105 ⟩
26. ⟨Formwork Carp ⟩	⟨125 ⟩
27. ⟨Steelfixer ⟩	⟨120 ⟩
28. ⟨Concretor ⟩	⟨100 ⟩
29. ⟨Conc. finisher ⟩	⟨105 ⟩
30. ⟨Scaffolder ⟩	⟨165 ⟩
31. ⟨Bricklayer ⟩	⟨105 ⟩
32. ⟨Bricklyr Labour⟩	⟨190 ⟩

RESOURCE SET — **CODE REFERENCES**

RESOURCE SET	CODE REFERENCES				
11. ⟨RC Stairs : TC/P ⟩	⟨27⟩	⟨26⟩	⟨28⟩	⟨ ⟩	⟨ ⟩
12. ⟨RC Stairs : H/ D ⟩	⟨27⟩	⟨26⟩	⟨28⟩	⟨ ⟩	⟨ ⟩
13. ⟨RC Foundations ⟩	⟨26⟩	⟨27⟩	⟨28⟩	⟨ ⟩	⟨ ⟩
14. ⟨Formwork ⟩	⟨26⟩	⟨ ⟩	⟨ ⟩	⟨ ⟩	⟨ ⟩
15. ⟨Steel fixing & concreting ⟩	⟨27⟩	⟨28⟩	⟨ ⟩	⟨ ⟩	⟨ ⟩
16. ⟨Structural steel erection ⟩	⟨60⟩	⟨ ⟩	⟨ ⟩	⟨ ⟩	⟨ ⟩
17. ⟨Asphalt roofing ⟩	⟨35⟩	⟨ ⟩	⟨ ⟩	⟨ ⟩	⟨ ⟩
18. ⟨Tiled roofing ⟩	⟨36⟩	⟨ ⟩	⟨ ⟩	⟨ ⟩	⟨ ⟩
19. ⟨Scaffolding ⟩	⟨30⟩	⟨ ⟩	⟨ ⟩	⟨ ⟩	⟨ ⟩
20. ⟨Brickwork ⟩	⟨31⟩	⟨32⟩	⟨90⟩	⟨ ⟩	⟨ ⟩

Figure 4.2(e) Resources library screen

the material cost, or if resources are involved the labour cost is calculated from the resource costs. The final duration and cost are then recorded. The program allows up to 200 such iterations, and the resulting different duration and cost estimates are displayed at the end.

Further facilities allow the results of the secondary level simulation and the effects of weather and preliminaries processing to be fed into the primary level to allow the fine detail to affect the overall result.

Example

Figure 4.2(a) shows the primary bar chart with 17 PWP bars for a contract extending over a period from 1980 to 1982. Figure 4.2(b) shows the secondary level bar chart for PWP 2 (substructure) containing 33 SWP bars. Figure 4.2(c) shows the cost screen for PWP 2 with the labour, material and total costs. Figure 4.2(d) shows the preliminaries schedule superimposed on the primary bar chart, the heavy dashed lines representing the duration of each preliminary category. Figure 4.2(e) shows part of the resource library which is used to indicate the man and machine power types needed for each activity, and Figure 4.2(f) shows the resources allocation screen which is used to indicate the quantity of each man or machine power type needed for each secondary level activity. Thus activity 5 (mass fill bases) requires three of resource set 24 (concreting only) which happens

```
PWP No. 2            ALLOCATE RESOURCES AND COSTS    Substructure (WB)
   ACTIVITY           RESOURCE SET       [TRADES]+<NUMBERS>
 4.[Exc Bases           ]  <26>*24<1 >:91<1 >:22<1 >:  <  >:  <  >:<   298>
 5.[Mass Fill Bases     ]  <24>*28<3 >:  <  >:  <  >:  <  >:  <  >:<   300>
 6.[Conc Lift Pit       ]  <13>*26<2 >:27<2 >:28<3 >:  <  >:  <  >:<   790>
 7.[Exc,drains & MH's(B)]  <26>*24<1 >:91<1 >:22<2 >:  <  >:  <  >:<   396>
 8.[Backfill & Blind    ]  <27>*12<1 >:13<2 >:  <  >:  <  >:  <  >:<   410>
 9.[Horiz. Bituthene    ]  <27>*12<1 >:13<3 >:  <  >:  <  >:  <  >:<   540>
10.[Protective Screed   ]  <27>*12<1 >:13<3 >:  <  >:  <  >:  <  >:<   540>
11.[Concrete (B) Slab   ]  <2 >*27<4 >:26<2 >:28<3 >:  <  >:  <  >:<  1030>
12.[Conc Retain W& Stair]  <8 >*27<2 >:26<2 >:28<3 >:  <  >:  <  >:<   790>
13.[Bituthene Retain W*1]  <27>*12<1 >:13<2 >:  <  >:  <  >:  <  >:<   410>
   RESOURCE SET                 CODE REFERENCES
18. [Tiled roofing          ][36]   [ ]     [ ]     [ ]     [ ]
19. [Scaffolding           ][30]   [ ]     [ ]     [ ]     [ ]
20. [Brickwork             ][31]   [32]    [90]    [ ]     [ ]
21. [Window fixing         ][37]   [38]    [ ]     [ ]     [ ]
22. [Plumbing             ][43]   [ ]     [ ]     [ ]     [ ]
23. [Mastic pointing      ][39]   [ ]     [ ]     [ ]     [ ]
24. [Concreting only      ][28]   [ ]     [ ]     [ ]     [ ]
25. [Excavation [large scale]][24] [92]    [ ]     [ ]     [ ]
26. [Excavation [small scale]][24] [91]    [22]    [ ]     [ ]
27. [General labouring    ][12]   [13]    [ ]     [ ]     [ ]
```

RESOURCE TOTALS

```
24. Grdwrk plantdvr: <2 >
92. Excavator      : <1 >
91. JCB            : <2 >
22. Grdwk Labourer : <2 >
28. Concretor      : <3 >
26. Formwork Carp  : <4 >
27. Steelfixer     : <4 >
12. Ganger         : <1 >
13. Labourer       : <3 >
```

Figure 4.2(f) Resources allocation screen

to comprise resource code 28 (concretor) at a unit cost of 100 (from Figure 4.2(f)). In other words the mass fill bases activity PWP 5 requires three concretors at 100 cost units each, i.e. 300 units.

Figure 4.2(g) shows the results obtained after 200 iterations of the simulation procedure. The average estimated cost and 95 % confidence limits can be read off the cumulative frequency curve at the appropriate 50, 2.5 and 97.5 percentage points, in this case £5 710 600 average with a range of £5 591 000 to £5 852 500, or £5 710 600 − 2.1 % + 2.5 %.

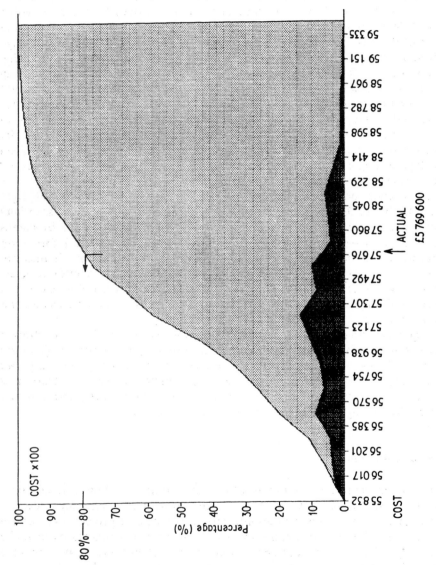

Figure 4.2(g) Simulation result, including weather and 16% project final cost

LEAD CONSULTANT EXPERT SYSTEM (ELSIE)

In 1982, a government committee under the chairmanship of John Alvey issued a report concerning Japanese research into 'Fifth Generation' computers, a general term covering several new computer technologies such as artificial intelligence and expert systems. As a result, the government created the Alvey Directorate, with a budget of £350m, to operate a five year collaborative research programme for the UK. One of the ensuing collaborative activities within the Intelligent Knowledge Based Systems (IKBS) sector was a 'community club', formed in 1986, comprising the members of the Royal Institution of Chartered Surveyors' Quantity Surveying Division (represented by a small group of practising chartered quantity surveyors) and a research team from the University of Salford (consisting of two knowledge engineers and a quantity surveyor). The original intention of the club was to examine the potential of expert systems in quantity surveying and this rapidly developed into an exercise in system building around the four core tasks involved in early stage strategic planning of construction projects — initial budget estimates, procurement choices, development appraisals, and duration estimates.

One of the major achievements of the 18 month research was the initial budget estimating facility of the system — a unique 'smart front end' to what is essentially a conventional approximate quantities estimating system. This is done by means of a knowledge based expert computer program which acts upon information more normally found in a project brief and automatically converts it into approximate quantity type information by making some reasonably intelligent assumptions. Like all expert systems, an important feature of ELSIE is that it can be interrogated about the assumptions it has made and these can be checked and amended where needed.

The major objective of the initial budgeting facility of ELSIE is to allow a budget to be set for the development of an office building, at an early stage in the project. In particular, it is designed to be used before scheme design drawings are available, but information such as sketch drawings can be utilised if available. It is designed to be used by semi-expert personnel, rather than novices, and in a way which will often complement rather than replace their activities.

Problems and limitations

In general the problems and limitations of the estimating part of the system are no different to those associated with any approximate quantities type technique — item rates need to be current and appropriate to the type of project involved (in ELSIE's case the latter is overcome by restricting present applications to office contracts only). The unique front end feature of the system however does involve some special considerations concerning the validity of the knowledge base.

The major problem concerning the validity of the knowledge base is whether and to what extent the design assumptions generated by the system correctly anticipate those of the designer. Innovative design or just simply changing fashions for example cannot be adequately handled by the system at present without intervention of the user. In general however the system is claimed to make design decisions of a kind that will be made up to five years after the start of the research, i.e. 1991. Obviously the validity of this claim remains to be seen.

Although in many cases the fashion element of design assumptions is made explicit so as to isolate the pertinent bits of knowledge base when modification is necessary, the facilities for user modifications are at present limited to unit costs and some certain default values within the database. Changes to the knowledge base generally can be made only by the software support team.

One further limitation of the system as a means of obtaining early stage estimates is that no information is produced to enable any kind of objective assessment of the reliability of the estimates. The only documented claim that has been made in this respect is that the estimates produced in testing the system (on about 40 building projects) were in general 'within about 5 % of that predicted by the expert quantity surveyor' (Brandon et al. 1988, p. 51), and in a number of cases better than the human expert. This figure should be treated with some caution however as it is likely that several of these tests were used in order to fine tune the program, and therefore are equivalent to the regression model derivation data. It is quite possible though, with this type of system, that the combination of human and computer 'expertise' working in tandem may well be greater than each working in isolation.

Applications

The initial budget facility of the ELSIE system, like most expert systems, operates in an interrogative manner asking the user a series of questions concerning the basic characteristics of the project from which the system can deduce a set of element type unit quantities and thence the likely building cost-price. The questions asked depend on the answers given to previous questions, but usually around 25 questions need to be answered. Once an answer is given the user has a chance to change some of the assumptions that have been made, before arriving at what is called the 'what-now' point.

Two basic options are available at the 'what-now' point, (1) make changes (in up to 150 items), (2) obtain reports. The facility to change the assumptions made, or the answers originally given to the questions, gives a comprehensive 'what-if' procedure, allowing the user to investigate the cost-price implications of various options and enabling the user to gain an impression of the sensitivity to design variation. The reports contain in varying degrees of detail cost breakdowns, explanations, and graphical representations of relationships such as quality with cost-price.

Example

This example illustrates the procedure used to produce a November 1986 estimate for a five storey speculative insurance company office building in the UK South East region for which information is available as given on page 111.

```
ELSIE BUDGET MODULE          Project:     0  ELSIE Test

CN30: Can you say what treatment the designers will give to the 'main' or
'front' elevation?

   If not absolutely sure, enter <U> (for Unknown at this stage) and you will
be asked about the image or quality required.  However, if you do happen to
know what is required, then enter one of the numbers below.

   1. Brick               5. PVC coated metal
   2. Brick/Stone mix     6. Exposed aggregate PC
   3. Natural Stone       7. GRP/GRC
   4. Prestigious Stone   8. Glazed Curtain Wall

Please note that stipulating one of 5..8 will automatically mean a framed
building.




                         (1 ..8 or U if not known,).. 1
```

Figure 4.3(a) Wall appearance input

```
ELSIE BUDGET MODULE          Project:     0  ELSIE Test

CN175: Can you specify the Number of Storeys the designers will give to the
building?

If not, then enter <U> for Unknown at this point, and a value will be
calculated from size of building or site.  If so, enter the number.

NOTE: This is the number of storeys at or above ground level, and does not
include basement levels.


   Please note that if you give a figure, any height restriction or plot ratio
limitations will be ignored.




                         (1 ..80 or U if not known,).. 5
```

Figure 4.3(b) Number of storeys input

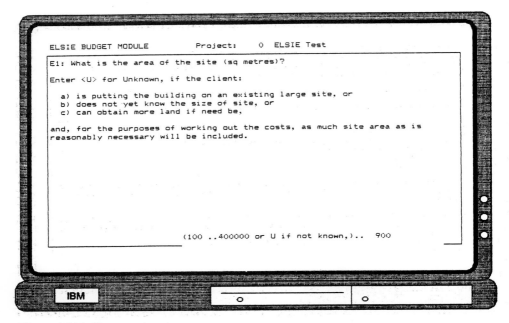

```
ELSIE BUDGET MODULE           Project:      0  ELSIE Test

E1: What is the area of the site (sq metres)?

Enter <U> for Unknown, if the client:

    a) is putting the building on an existing large site, or
    b) does not yet know the size of site, or
    c) can obtain more land if need be,

and, for the purposes of working out the costs, as much site area as is
reasonably necessary will be included.




                              (100 ..400000 or U if not known,)..  900
```

Figure 4.3(c) Site area input

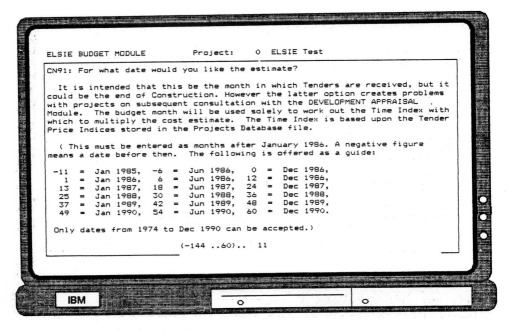

```
ELSIE BUDGET MODULE           Project:      0  ELSIE Test

CN91: For what date would you like the estimate?

    It is intended that this be the month in which Tenders are received, but it
could be the end of Construction. However the latter option creates problems
with projects on subsequent consultation with the DEVELOPMENT APPRAISAL  .
Module.  The budget month will be used solely to work out the Time Index with
which to multiply the cost estimate.  The Time Index is based upon the Tender
Price Indices stored in the Projects Database file.

    ( This must be entered as months after January 1986. A negative figure
means a date before then.  The following is offered as a guide:

   -11  =  Jan 1985,   -6  =  Jun 1986,    0  =  Dec 1986,
     1  =  Jan 1986,    6  =  Jun 1986,   12  =  Dec 1986,
    13  =  Jan 1987,   18  =  Jun 1987,   24  =  Dec 1987,
    25  =  Jan 1988,   30  =  Jun 1988,   36  =  Dec 1988,
    37  =  Jan 1989,   42  =  Jun 1989,   48  =  Dec 1989,
    49  =  Jan 1990,   54  =  Jun 1990,   60  =  Dec 1990.

Only dates from 1974 to Dec 1990 can be accepted.)

                         (-144 ..60)..  11
```

Figure 4.3(d) Tender date input

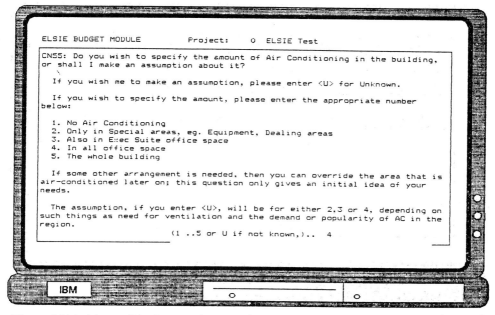

```
ELSIE BUDGET MODULE          Project:    0  ELSIE Test

CN55: Do you wish to specify the amount of Air Conditioning in the building,
or shall I make an assumption about it?

   If you wish me to make an assumption, please enter <U> for Unknown.

   If you wish to specify the amount, please enter the appropriate number
below:

   1. No Air Conditioning
   2. Only in Special areas, eg. Equipment, Dealing areas
   3. Also in Exec Suite office space
   4. In all office space
   5. The whole building

   If some other arrangement is needed, then you can override the area that is
air-conditioned later on; this question only gives an initial idea of your
needs.

   The assumption, if you enter <U>, will be for either 2,3 or 4, depending on
such things as need for ventilation and the demand or popularity of AC in the
region.
                       (1 ..5 or U if not known,).. 4
```

```
IBM                              o                    o
```

Figure 4.3(e) Air-conditioning requirements input

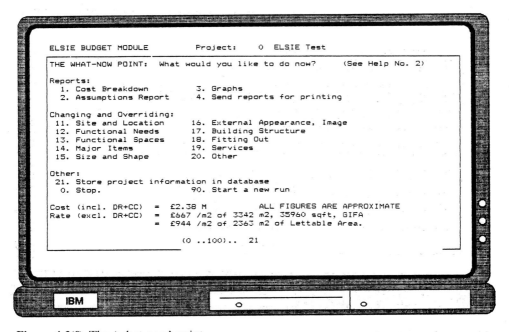

```
ELSIE BUDGET MODULE          Project:    0  ELSIE Test

THE WHAT-NOW POINT:  What would you like to do now?     (See Help No. 2)

Reports:
   1. Cost Breakdown       3. Graphs
   2. Assumptions Report   4. Send reports for printing

Changing and Overriding:
   11. Site and Location    16. External Appearance, Image
   12. Functional Needs     17. Building Structure
   13. Functional Spaces    18. Fitting Out
   14. Major Items          19. Services
   15. Size and Shape       20. Other

Other:
   21. Store project information in database
    0. Stop.              90. Start a new run

Cost (incl. DR+CC)  =  £2.38 M        ALL FIGURES ARE APPROXIMATE
Rate (excl. DR+CC)  =  £667 /m2 of 3342 m2, 35960 sqft, GIFA
                    =  £944 /m2 of 2363 m2 of Lettable Area.

                       (0 ..100).. 21
```

```
IBM                              o                    o
```

Figure 4.3(f) The 'what-now' point

REPORT ON PROJECT

INPUT INFORMATION

('X ..Y' below means the information was not asked for)

No.	Item	Value
1	type of client	4 Speculative
2	sector	4 Insurance
3	office status	1 ..5 Sole Office to Main HQ
4	office function	1 ..3
10	size derivation	3
11	office staff needed	10 ..10000
13	office space wanted	100 ..500 000 m^2
170	GIFA wanted	3342 m^2
175	no. of storeys	5
14	max. storeys	0 ..80
21	basement needed?	False
176	basement area	0 m^2
20	no. of cars	5
30	wall appearance	1 : Brick
191	wall detailing	—————
32	image	1 ..7
50	level of fitting out	2
51	open plan %	100%
52	need for large spaces	0
53	column free?	False
54	airtightness	0 ..10
55	AC where?	4: Office Space
150	AC type	5: VAV
70	region	6 South-East region (excl. London)
71	type of locality	1 ..5 City centre to Rural/village
72	neighbour distance	4
72	noise level	10
179	site area	900 m^2
81	site known?	True
82	site difficulty	7
83	state of site	2
84	ground bearing capcty	1 ..4
85	special foundation?	True
86	amount of rock	0
87	water problems	0
88	slope	1
89	can build on rock	Unknown
90	construction months	15 months
91	start month	Nov 1986

Figure 4.3(g) Answer summary sheet for ELSIE run

Size Derivation
The size of the building (GIFA) was directly specified.
The height of the building was specified as 5 storeys.
The size of the site was specified directly.

Main Results
Approximately, the construction costs will be £2.38 m, including Design Reserve and construction contingencies of 6.5%.

Without these contingencies the cost is approximately £2 230 468, and gives average rates over the gross internal floor Area of:

£667 per square metre over 3342 m^2
£ 62 per square foot 35960 ft^2

This price is based on conventional procurement path with competitive tendering. It includes preliminaries of around 17%. These have been increased above a typical of 12% to take into account limitations on the construction process due to site constraints.
Location and tender price indices have been included as follows:
    Location: 1.00 for south-east region (excl. London)
                   (1.0 = south east region)
    Budget date: 1.03 for Nov 1986 (1.0 = Jan 1986)

You can alter the budget date to be start or end of construction period, or at any other time, to suit your requirements.

Elemental Breakdown		%	0 50	£/m^2	£
1	Substructure	4		28	92 346
	Basement	0		0	0
2A	Frame	8		56	187 053
B	Upper floors	5		32	107 279
C	Roof	3		21	69 018
D	Stairs	3		19	63 460
E	External walling	7		44	145 512
F	Windows + ext doors	7		47	158 539
G	Internal walls / doors	2		11	37 341
3	Finishes	8		51	169 533
4	Fitting and furnishings	1		6	20 583
5F	Heating and ventilation	17		114	380 378
H	Electrical	9		59	198 607
J	Lifts	3		20	67 189
M	Special installations	1		4	14 411
	Other services and EWIC	4		29	95 866
6	External services/work	2		13	44 174
7	Prelims	17		113	379 180
Total (less contingencies)		100		667	2 230 468

Figure 4.3(h) Main summary report ELSIE

Site Usage

	Metric	Acres
Landscaped area	122	0.0
Hard area (e.g. car park)	110	0.0
External site area	232	
Building footprint	668	
Site fit	0	0
Site area	900	0.2
Total car parking area needed	110	0.0
No. of car spaces	5	

(The assumed paved area does not allow for any paved concourses nor approach roads; if you wish to include these then override the paved area.)

Size and Shape
The building is reckoned to be simple rectangle in plan shape, 5 storeys high. The major areas are as follows:

	Metric	Imperial
Ground floor area	668	7 192
Upper floor area	2674	28 768
Basement area	0	0
Atrium area	0	0
Gross internal floor area	3342	35 960
Average floor–floor height	3.8	12.6
Width of building	13.6	45
Overall Length	49	163
Wall/floor ratio	0.72	

The average plan area for upper floors is $668\,m^2$.

External Appearance
The AA quality and external complexity factors are:

Aesthetic/amenity quality   _____
Complexity of form   _____
        of walls _____
        of roof   _____

The 'main' elevation may be something like brick on frame, with a window-to-wall ratio of 0.33.

Figure 4.3(i) Example of one sheet of ELSIE assumptions report

(1) 3342 m² gross floor area, 100 % lettable
(2) Brick elevation
(3) No basement
(4) 15 months contract period
(5) 900 m² level town site with access problems
(6) VAV air-conditioning in offices only
(7) Very noisy location
(8) 5 car parking spaces

Figures 4.3 (a)−(e) show some of the screens involving the question and answer brief elicitation phase. Figure 4.3(f) gives the resulting cost-price estimate generated by the system (£2.38m) at the 'what now' stage. Figures 4.3 (g)−(i) show the first three pages of the explanations report (12 pages in total were produced), including an elemental breakdown of the estimate.

BIDDING MODEL DEBIASER

Although not strictly estimating techniques, estimate debiasers constitute a collection of very new 'back end' techniques, still in the research phase, aimed at improving or fine tuning estimates generated by other techniques. Three types of debiasers are under current development:

(1) Regression debiasers, which are identical to the usual regression estimating techniques except that the estimate is specifically included in the predictor variable list.
(2) Control chart debiasers, using dynamic time series detrending techniques to detect real time biasing of recent estimates.
(3) Bidding model debiasers, which utilise bidding theory to assess bias and reliability in estimates.

This section describes one of the bidding model debiasers being developed by the authors.

The purpose of bidding models is to enable a bidder to assess the best mark up value to use in an auction given some information concerning the likely bids to be entered by his competitors. This information can range from a knowledge of the identity and past bids of all the competitors in the auction through to virtually no information at all. Many formulations have been proposed to model this situation. The model adopted here is the multivariate model (Appendix B). As the probability of entering the lowest bid with a given bid is exactly the same as the probability that the lowest bid will be less than an estimate of the lowest bid, the model can be applied without modification.

Problems and limitations

The problems and limitations are similar to those of the regression estimating approach in concerning model assumptions and data limitations.

Model assumptions

Three major assumptions concerning the validity of equation (B.1) need to be addressed

(1) the log normal assumption;
(2) the independence assumption; and
(3) the consistency assumption.

The first of these assumptions is dealt with in Appendix B.

Violations of the independence assumption may have much more severe consequences. Independence is however very difficult to establish with data of this kind, and possible violations tend to be ignored for this reason. It is probably now a truism that possible lack of independence is the reason for many models of this kind failing to achieve commercial status.

The consistency assumption, i.e. that bidders behave in much the same way irrespective of the type, size and other characteristics of contracts, is a reflection on the simplicity of the model (of which the independence assumption is a special case). Some research is currently proceeding on this aspect.

Data limitations

Although most bidding models have very heavy demands on data, particularly on the frequency with which certain specified bidders compete against each other, the multivariate model is relatively undemanding in this respect.

The data consist of any previous designer's estimates together with contractors' bids. It is not necessary that designer's estimates are available for all the contracts in the data, nor that the designer's estimates and the specified contractors' bids are recorded for the same contract. All that is necessary is an indirect link between the designer and the specified bidders. For example, if contractors A, B, and C are bidding for a new contract, it is not important that the designer has produced estimates for previous contracts on which A, B, and C have entered bids, nor that any of the people involved have bid against each other before. All that is required is that all the people involved have bid at least once against another bidder who has bid at least once against the current competitors, or have bid at least once against another bidder who has bid at least once against another bidder who has bid at least once against the current competitors, etc.

There is a price to pay however for the relative lack of data restriction, reflected in the independence and consistency assumptions mentioned above.

Applications

The bidding model debiaser involves three stages

• data preparation and entry;
• estimation of parameters; and
• calculation of probability of lowest bid values.

Data preparation and entry

The data consists of all available designers' estimates, bids and associated bidders' names for a set of historical construction contract auctions. These are entered into an auction database in the form of a contract number, designer's estimate value/ designer code, and bid value/bidder code.

Estimation of parameters

The computer program automatically calculates the required model parameters from equations (B.2) and (B.3) and stores the results in a computer file. This operation is only necessary when new historical auction data is entered into the auction data base, and takes a few seconds of computer time.

Calculation of the probability of the lowest bid values

The computer program automatically calculates the unbiased estimated probability values for each of a sequence of m values and plots the resulting curve in terms of P'. Further probability estimates are then obtained via estimates of μ and σ^2 obtained by stochastic simulation, each iteration generating a different probability value. These additional values provide an indication of the variability of the probability estimates and are plotted as points on the graph. The resulting graph therefore enables the user to gain an impression not only of the reliability of the debiased estimate but also of the reliability of the reliability!

Example

This example contains data collected from a London building contractor (coded number 304) for an incomplete series of construction contracts auctioned during a 12 month period in the early 1980s (Table 4.4). For the purposes of this example, the bids entered by bidder 304 are treated as designer's estimates.

The program requests the value of the estimate for the new contract together with the identity of the bidders. In this example the estimate of the new contract is £3m and the bidders are code 55, 73, 134, 150 and 154. The program then automatically proceeds as follows:

(i) transforms the data to the log values $y_{ik} = \log (x_{ik} - \lambda x_{(1)k})$, in this case $\lambda = 0.6$;

(ii) calculates the required model parameter estimates, α_i, β_k, and s^2_i;

(iii) calculates the probability of code 304 'underbidding' the other bidders with $m = -0.70, -0.69$, etc. by substituting α_1 and s_i for μi and σ_1 in equation (B.6);

(iv) generates a value for μ_{304}, σ^2_{304}, μ_{55}, σ^2_{55}, μ_{73}, σ^2_{73}, μ_{134}, σ^2_{134}, μ_{150}, σ^2_{150}, μ_{154}, σ^2_{154}, by stochastic simulation;

(v) calculates the probability of code 304 'underbidding' the other bidders with m values obtained by stochastic simulation;

(vi) repeats (iv) and (v) 600 times;

(vii) detransforms the m values to P' by equation (B.10) and plots the curve resulting from step (iii) and points resulting from steps (iv) to (vi).

Table 4.4 Data for incomplete series of construction contracts

PROJ	BID	BDR	BID	BDR	BID	BDR	BID	BDR	BID	BDR	BID	BDR	BID	BDR	BID	BDR	BID	BDR
1	1454515.	150	1514865.	55	1475398.	304	1468775.	304	1447867.	134	1457977.	154	1865545.	73		93		360
2	535608.	304	502042.	291	529744.	154	516376.	157										
3	1333142.	75	1331156.	217	1366863.	304	1266892.	281	1276787.	115	1277652.	93						
4	696743.	304	696972.	292	701062.	237	637815.	79	637815.	361	697826.	157						
5	404110.	55	422297.	304	413224.	97	389196.	117	417489.	362	389848.	157						
6	2116877.	134	2169966.	99	2187991.	293	2161120.	304	2198855.	221	2296108.	137	2165611.	8	2153344.	117	2133608.	294
7	3065742.	304	3119689.	150	3141641.	170	3153800.	247	3161120.	304	3249927.	134	3269768.	191	3335993.	55		187
8	7925257.	221	7351929.	304	7374650.	247	6900000.	20										
9	871520.	118	899935.	137	902378.	304	914393.	304	950737.	83	996483.	221						
10	1063337.	304	1154023.	251	1102272.	173	1079657.	201										
11	1759614.	154	1792123.	281	1838532.	157	1918066.	170	1947733.	304	1784215.	304						
12	1126816.	304	1146398.	201	1169795.	154	1227294.	24	1312527.	280	1399472.	280						
13	698005.	304	625501.	268	630288.	308	666545.	308										
14	588810.	364	584833.	365	639229.	79	646341.	145	682802.	304	691474.	304						
15	1429218.	303	1493849.	291	1511033.	304	1521628.	304	1526377.	366	1717715.	55						
16	842319.	6	870894.	304	883617.	185												
17	284947.	367	292692.	356	294494.	368	303700.	152	307282.	85	313203.	134	315727.	369	333597.	118	334353.	370
	348969.	304																
18	461444.	150	483862.	304	482241.	308	447021.	308	493417.	154	455480.	311						
19	2858191.	280	2947007.	371	2950723.	134	2999999.	304	3093582.	60	3099528.	6	3278229.	266	3335198.	170	3333793.	55
20	7831865.	276	7837276.	304	7859122.	256	7904172.	55	8047230.	152	8145323.	293	8279564.	117	8657685.	134		
21	3971051.	55	3854074.	304	4724785.	372	3955009.	154	3944772.	373	3731543.	79	4001188.	237				
22	573485.	292	596737.	292	597730.	134	613528.	134	615015.	304	621223.	170						
23	1610942.	304	1623447.	163	1646286.	173	1663742.	268	1700000.	152								
24	1196036.	64	1199328.	374	1208837.	187	1226589.	304	1262082.	291	1271000.	291	1295954.	170	1302161.	254		
25	2636397.	137	2654728.	150	2673906.	187	2685122.	55	2762123.	304	2845567.	304	540814.	304				
26	469663.	24	476784.	268	485870.	286	486485.	55	504026.	122	529468.	263						
27	1526553.	201	1533719.	152	1698797.	148	1876612.	304										
28	2106139.	201	2175928.	304	2210065.	308	2223710.	308	2255246.	221	2296623.	117	2331830.	266				
29	499888.	102	559596.	55	592026.	217	602042.	170	608957.	304	619065.	134						
30	2639525.	304	2842407.	308	2874130.	280	2861665.	280	2736300.	152	2770720.	256						
31	732572.	304	599429.	365	623906.	145	691759.	79	744332.	154	607065.	364						
32	546641.	134	539565.	268	608742.	55	538382.	24	599934.	170	559351.	304						
33	792966.	221	811788.	99	819971.	308	847621.	55	847892.	137	853793.	304	2325900.	304				
34	2085151.	152	2130217.	107	2150583.	280	2203956.	280	2219653.	137	2241687.	154						
35	821617.	268	844579.	115	848459.	303	871927.	304	872215.	106	935745.	375						
36	792474.	304	747374.	24	778559.	217	743788.	252	808345.	268	835465.	170						
37	7279854.	304	7650271.	60	7702448.	308	6631664.	308	7089879.	193	7230120.	170	6986341.	247	7143710.	191	6794553.	266
38	592096.	348	573997.	150	518613.	217	508985.	217	544480.	121								
39	538600.	154	567031.	377	621365.	378	699839.	378	825451.	72	991468.	190	1001254.	304				
40	2087946.	247	2104017.	276	2183122.	186	2205359.	186	2212382.	280	2267987.	112	2232476.	221	2400000.	294		
41	1503739.	191	1536654.	24	1576905.	304	1583595.	154	1616432.	294	1704995.	157						
42	3624453.	157	3694803.	221	3732133.	304	3751115.	304	3773967.	193	3866339.	55	3922937.	134	4122448.	281		
43	629164.	304	695284.	173	723315.	311	729305.	266	743578.	304	768189.	379						
44	2252833.	163	2264310.	24	2274380.	112	2323385.	304	2384494.	55								
45	1202916.	217	1268733.	55	1291365.	221	1294986.	221										
46	2968891.	286	2772626.	280	2822857.	186	2972189.	134	2821600.	276	2857275.	304	2793000.	221				
47	1398400.	294	1401500.	152	1427140.	237	1436804.	237	1453070.	301	1511643.	55	1591986.	371	1665760.	83		
48	698161.	31	709676.	291	758565.	291	789355.	304	789926.	134	842684.	55	751677.	252				
49	248733.	293	251007.	291	251415.	252	261286.	380	264933.	304								
50	358840.	317	362370.	217	386983.	304	421797.	381	456272.	154								
51	527692.	311	570874.	311	588854.	75	609221.	173	636451.	308	694297.	304						

The resulting graph is shown in Figure 4.4. The graph is interpreted by drawing a horizontal line at the 50, 2.5 and 97.5 percentage probability points across to the curve and thence down to the estimate axis as shown to obtain the unbiased estimate (£2 827 300) and 95 % confidence limits (£2 566 700 to £3 138 000, i.e. £2 827 300 + 10.99 % − 9.22 %) due to the variability of the designer's estimates and contractors' bids *as predicted by the model*.

The surrounding points indicate the effect of the size of the database on the reliability of the parameter estimates in the model − the true curve will be contained somewhere within these points. With the small amount of data used in this example, the points are quite widespread. The existence of a larger database should have the effect of decreasing this spread.

CONCLUSION − THE FUTURE?

Contract price forecasting techniques clearly comprise a large topic area worthy of a book in its own right. As Table 4.1 indicates, the field is rapidly developing out of the older deterministic approaches into methods which specifically accommodate the inherent variability and uncertainties involved in forecasting the price of construction work. One result of this is that the traditional distinction between 'early stage' or 'conceptual' estimating and 'later stage' or 'detailed' estimating is

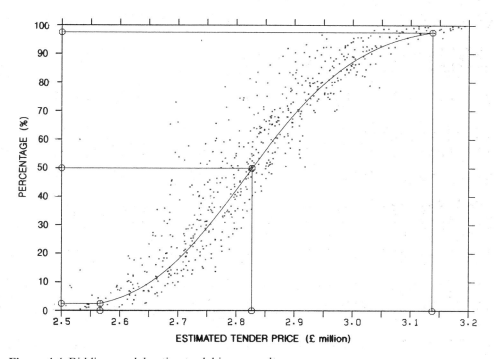

Figure 4.4 Bidding model estimate debiaser results

being replaced by the more fundamental distinctions concerning the reliability of forecasts and their components — items, quantities and rates. This has focussed attention on the means of modelling and predicting reliability — statistically for simplicity and stochastically for complexity. The construction project simulator, for instance, contains stochastic elements for item quantities, whilst risk estimating utilises both statistical and stochastical techniques. Little has been done to treat the items themselves in this way, although some relatively new quantity generation systems, such as ELSIE, are clearly capable of extension. The logical conclusion of these approaches will be a technique which combines all three elements of the forecasting equation into the same non-deterministic, item, quantity, rate, (NDIQR) system.

Although a somewhat daunting prospect for practitioners, the development of NDIQR systems will mark a new and exciting phase in the evolution of construction price forecasting systems generally. Firstly, current deterministic requirements will still be accommodated as deterministic forecasts are simply a special case for a non-deterministic system. A rate with mean say £5 and standard deviation £0 is effectively a deterministic rate. Also the 'best guess' of a non-deterministic system is a deterministic answer. Thus the range of forecasts provided by a non-deterministic system can be regarded as secondary information to the deterministic forecast, to be divulged or not as the user wishes. Secondly, NDIQR systems will allow forecasts to be made at any stage of the design process. Treating the items themselves as random variables, for example, means that the standard deviation simply reduces as we become more certain that the item will be appropriate. Thus a BQ PRICING NDIQR system will commence with a notional bill of quantities that will gradually firm up as the design progresses. Thirdly and perhaps most importantly, the reliability measures provided by NDIQR systems will enable comparisons to be made between alternative systems. For example, if a practice uses several systems to provide price forecasts for the same contract and obtains the following results:

System	Forecast range
A	£3.2m to £3.9m
B	£3.5m to £4.0m
C	£3.7m to £4.5m

we may select the inner range of these three systems, i.e. £3.7m to £3.9m, to be the best range of forecasts.

APPENDIX A: RELIABILITY OF REGRESSION FORECASTS

95% confidence limits

This is obtained from the standard error of the forecast SE (Y) where

$$SE(Y)^2 = S^2 (1 + 1/N) + \sum_{i=1}^{k} (x_i - \bar{x}_i)^2 S^2 c^{ii} + \sum_{i=1}^{k-1} \sum_{j>i}^{k} 2(x_i - \bar{x}_i)(x_j - \bar{x}_j) S^2 c^{ij} \quad \textbf{(A.1)}$$

where S^2 is the mean square of the residuals, N is the number of previous cases, $\bar{x}_1$, $\bar{x}_2$, $\bar{x}_3$, etc. are the mean values of the independent variables x_1, x_2, x_3, etc., and $S^2 c^{ij}$ are the variance−covariances of the regression coefficients.

The 95 % confidence limits are then approximated by

$$\pm\, t_{(0.025)\ (N-n-1)}\ SE(Y) \tag{A.2}$$

where $t_{(0.025)\ (N-n-1)}$ is obtained from the students t distribution tabulated in most elementary statistical texts.

Coefficient of variation

The coefficient of variation, cv, can be obtained from the distribution of jackknife deleted residuals as follows

$$cv = 100 S_d / (x_p - \bar{x}_d) \tag{A.3}$$

where x_d and S_d represent the mean and standard deviation of the deleted residuals and x_p represents the mean prediction. This statistic, though not conventionally used in regression applications, has the advantage of being directly comparable with the variability measures associated with other techniques and studies.

Reliability of forecast for Salford Offices example

The standard error of forecast is obtained from equation (A.1) as

$$SE(Y)^2 = S^2\,(1 + 1/N) + (x_1 - \bar{x}_1)^2 S^2 c^{11} + (x_2 - \bar{x}_2)^2 S^2 c^{22} + (x_3 - \bar{x}_3)^2 S^2 c^{33} +$$
$$2\,(x_1 - \bar{x}_1)\,(x_2 - \bar{x}_2) S^2 c^{12} + 2\,(x_1 - \bar{x}_1)\,(x_3 - \bar{x}_3) S^2 c^{13} +$$
$$2\,(x_2 - \bar{x}_2)\,(\bar{x}_3 - \bar{x}_3) S^2 c^{23} \tag{A.4}$$

where x_1, x_2, x_3, represent the log values of bidders GFA and period respectively. The regression output (Figures 4.1(a) and (b)) at step 3 shows that

$S^2 = 0.07648$, $N = 27$, $\bar{x}_1 = 1.703$, $\bar{x}_2 = 7.103$, $\bar{x}_3 = 2.336$, $S^2 c^{11} = 0.01211$, $S^2 c^{22} = 0.00858$, $S^2 c^{33} = 0.07333$, $S^2 c^{12} = -0.00390$, $S^2 c^{13} = 0.01126$, $S^2 c^{23} = -0.02226$.

Substituting into the above equation gives:

$$SE(Y)^2 = 0.07648\,(1 + 1/27) + (\log 6 - 1.703)^2 0.1211 + (\log 6000 - 7.103)^2$$
$$0.00859 + (\log 18 - 2.336)^2 0.07333 - 2\,(\log 6 - 1.703)$$
$$(\log 6000 - 7.103)\,0.00390 + 2\,(\log 6 - 1.703)\,(\log 18 - 2.336)$$
$$0.01126 - 2\,(\log 6000 - 7.103)\,(\log 18 - 2.336)\,0.2226$$

$$= 0.0793 + 0.0001 + 0.0219 + 0.0225 - 0.0011 + 0.0011 - 0.0197$$
$$= 0.0844$$

so $SE(Y) = 0.2905$.

The 95% confidence limits for the standardised log forecast is then, from equation (A.2)

$$Y \pm t_{(0.025)\ (27-3-1)}\ 0.2905 = 1.0609 \pm 2.069 \times 0.2905$$
$$= 1.0609 \pm 0.6010 \text{ i.e. } 0.4599 \text{ to } 1.6619$$

which is 1.5839 to 5.2693 for the standardised forecast, and £2 565 918 to £8 536 266 for the Salford January 1989 contract.

The coefficient of variation is obtained by dividing the standard deviation of the forecast error ratios by their mean. As a log model has been used, the antilog of the deleted residual will give the required ratio. At step 3 the standard deviation is 0.3339 and the mean 1.0419, giving a coefficient of variation of 32.05%.

Some statistical packages make the calculations easier by providing a direct forecast for the new contract. In this example we have entered the new contract into the data base with a dummy value of £0.01 but excluding it from the model by a special select instruction. As Figure 4.1(c) shows, the correct forecast of 1.0609 is obtained. This package also gives the standardised deleted residual SDRESID, which is the deleted residual DRESID divided by its standard error. The standard error of the forecast can therefore be obtained very quickly as

$$SE(Y) = DRESID/SDRESID$$
$$= -14.3656/-48.4444 = 0.2905 \qquad \textbf{(A.5)}$$

The residual statistics (Figure 4.1 (d)) also enable a quick approximation of the coefficient of variation to be made by

$$cv = 100 \times \text{std dev DRESID}/(\text{mean PRED} - \text{mean DRESID})$$
$$= 100 \times 0.3751/(1.0880 - 0.0187) = 35.08 \qquad \textbf{(A.6)}$$

Although this figure is the coefficient of variation of the log values, it is proportional to the raw coefficient of variation and therefore indicative of the relative values. Also, as the model becomes more reliable, then the coefficient of variation of the logs values becomes closer to the coefficient of variation of the raw values

APPENDIX B: THE MULTIVARIATE BIDDING MODEL DEBIASER

The multivariate bidding model is

$$\log(x_{ik}) = y_{ik} \sim N(\mu_i + \mu_k, \sigma^2_i) \qquad \textbf{(B.1)}$$

where x_{ik} is bidder i's bid (i = 1, 2, ..., r) entered for auction k (k = 1, 2, ..., c), the log of which is normally distributed with mean μ_i, a bidder location parameter, plus μ_k, an auction size datum parameter, and with a unique variance parameter σ^2_i for each bidder. These parameters can be estimated quite easily by an iterative procedure solving

$$\beta_k = \hat{\mu}_k = \sum_{i=1}^{r} \delta_{ik} \, (y_{ik} - \alpha_i)/n_i \qquad \textbf{(B.2)}$$

$$\alpha_i = \hat{\mu}_i = \sum_{k=1}^{c} \delta_{ik} \, (y_{ik} - \beta_k)/n_i \qquad \textbf{(B.3)}$$

and thence

$$S^2_i = \hat{\sigma}^2_i = \sum_{k=1}^{c} \delta_{ik} \, (y_{ik} - \alpha_i - \beta_k)^2 / \left\{ (n_i - 1) \left(1 - \frac{c-1}{N-r} \right) \right\} \quad \text{for } n_i > 1 \quad \textbf{(B.4)}$$

$$= \sum_{k=1}^{c} \delta_{ik}(y_{ik} - \alpha_i - \beta_k)^2/(N - c - r + 1) \qquad \text{for } n_i = 1 \quad \textbf{(B.5)}$$

where Kronecker's $\delta_{ik} = 1$ if bidder i bids for auction k
$\qquad\qquad\qquad = 0$ if bidder i does not bid for auction k

$n_i = \sum_{k=1}^{c} \delta_{ik}$, the number of bids made by bidder i

$N = \sum_{k=1}^{c} n_i$, the total number of bids by all bidders

The probability of a bidder, say i = 1, underbidding a set of specified competitors on a contract is then given by

$\Pr(y_1 + m < y_i, \, i \neq 1) =$

$$\int_{-\infty}^{\infty} \{(2\pi)^{\frac{1}{2}}\}^{-1} \exp\left(-\tfrac{1}{2}y_1^2\right) . \left\{ \prod_{i=2}^{n} \int_{y_i = (\sigma_1 y_1 + \mu_1 + m - \mu_i)\sigma_i^{-1}}^{\infty} \{(2\pi)^{\frac{1}{2}}\}^{-1} \exp\left(-\tfrac{1}{2}y_i^2\right) dy_i \right\} dy_1 \quad \textbf{(B.6)}$$

where m is a 'decision' constant used by bidder 1 to bring about a desired probability state (the usual approach is to use bidder 1's cost estimates in the analysis in preference to his bids on the assumption that the m value will reasonably approximate his likely profit should he acquire the contract). To observe the effects of the limited accuracy of the parameter estimates due to the sample size, values for μ and σ are obtained from α and s^2 from their sampling distributions

$$\mu_i \sim N \, (\alpha_i, \, s^2_i/n_i) \qquad \textbf{(B.7)}$$

$$\sigma^2_i(n_i - 1)s^2_i \sim \chi^2(n_i - 1) \qquad\qquad \text{for } n_i > 1 \qquad \textbf{(B.8)}$$

$$\sigma^2_i(N - c - r + 1)/s^2_i \sim \chi^2(N - c - r + 1) \qquad \text{for } n_i = 1 \qquad \textbf{(B.9)}$$

Now if we substitute the designer's estimator for bidder 1 in the above formulation, the value of m which results in a probability of 0.5 represents the bias in his (log) estimate and thus the amount that needs to be added to his estimates to give an unbiased estimate of the lowest bid. Also the values of m which result in probabilities of 0.025 and 0.975, will give the 95% confidence limits.

The log normal assumption has been tested with three sets of UK construction contract bidding data indicating that a three parameter log normal model may be more appropriate than the general two parameter model proposed here. The modifications necessary to convert the formulation are quite straightforward however, involving a prior transformation of the data before applying the iterative procedure. The resulting m values have however to be detransformed before plotting the final probability graph. Thus, for data transformed by $y_{ik} = \log (x_{ik} - \lambda x_{(1)k})$, the unbiased estimate P′ is given by

$$P' = \frac{\lambda P e^{\omega}(1 - e^m)}{1 - \lambda + \lambda e^{\omega}} + Pe^m \qquad (\omega = m \,|\, Pr = 0.5) \qquad \textbf{(B.10)}$$

BIBLIOGRAPHY

Barnes, N. M. L. (1971) *The design and use of experimental bills of quantities for civil engineering contracts.* PhD thesis, University of Manchester Institute of Science and Technology.

Bennett, J. and Ormerod, R. N. (1984) Simulation applied to construction projects. *Construction Management and Economics*, **2**, 225–263.

Brandon, P. S., Basden, A., Hamilton, I. W., Stockley, J. E. (1988) *Application of Expert Systems to Quantity Surveying: the Strategic Planning of Construction Projects.* London: Surveyors Publications, RICS.

Fine, B. (1980) *Construction Management Laboratory.* Fine, Curtis and Gross.

Gilmore, J. and Skitmore, M. (1989) A new approach to early stage estimating. *Chartered Quantity Surveyor*, May.

Lu Qian (1988) Cost estimation based on the theory of fuzzy sets and predictive techniques. *Construction Contracting in China.* Hong Kong Polytechnic.

Ross, E. (1983) *A Database and Computer System for Tender Price Prediction by Approximate Quantities.* MSc project report, Loughborough University of Technology.

Southgate, A. (1988) Cost planning — a new approach. *Chartered Quantity Surveyor*, November, p. 35–36.

Chapter 5

Risk Analysis

ROGER FLANAGAN, *Professor of Construction Management,
University of Reading* and SUSAN STEVENS, *Crest Estates,
Weybridge*

INTRODUCTION

The construction industry is subject to more risk and uncertainty than probably
any other industry. Getting a project from the inception stage through to completion
involves a vast number of people with differing skills. Buildings tend to be
bespoke and each new project involves new design and construction problems
that have to be overcome.

Risk can manifest itself in numerous ways, varying over time and across activities. Essentially it stems from uncertainty which, in turn, is caused by a lack of
information. The environment within which decision-making takes place can be
divided into three parts:

- certainty;
- risk;
- uncertainty.

Certainty exists only when one can specify what will happen during the period of
time covered by the decision; that does not happen very often in the construction
industry. An important source of bad decisions is fairly often *illusions of certainty*.
Most people who earn their living in the construction industry are optimists.

There is a difference between risk and uncertainty. A decision is made under
risk when a decision-maker can assess, either intuitively or rationally, the probability of a particular event occurring, the probabilities of the event being based
upon historical data or 'experience'. For instance, the quantity surveyor frequently
has to establish the budget price for the foundations to a building without
knowing the ground conditions or the loading for the building. There is an
element of risk in forecasting the budget price, but past data and experience tell
the quantity surveyor that it can be achieved within the budget price with at least
some degree of certainty.

Uncertainty, by contrast, might be defined as a situation in which there are no
historic data or history relating to the situation being considered by the quantity
surveyor. An example would be the building of a high technology building which
is to be heated by a new innovative technique using electronic pulses to heat the

fabric. If such a structure had not been built before, the quantity surveyor would have no historic data on which to base the decision on price.

The differences between risk and uncertainty are close, and for convenience the construction industry uses the term risk to encompass both risk and uncertainty. Typical risks on a construction project include:

Source + risk

- failure to complete within the stipulated design and construction time;
- failure to obtain the expected outline planning, detailed planning or building regulation approvals within the time allowed in the design programme;
- unforeseen adverse ground conditions delaying the project;
- exceptionally inclement weather delaying the project;
- strike by the labour force causing disruption to the construction programme;
- unexpected price rises for labour and materials;
- failure to let a speculative development to a tenant upon completion;
- an accident to an operative on site causing physical injury;
- latent defects occurring in the structure through poor workmanship;
- *force majeure* (flood, earthquake, etc.) disrupting the work;
- a claim from the contractor for loss and expense caused by the late production of design details by the design team;
- failure to complete the project within the client's budget allowance.

It is important to distinguish the sources of risk from their effects. Ultimately, all risk encountered on a project is related to one or more of the following:

- failure to keep within the cost budget/forecast/estimate/tender;
- failure to keep within the time stipulated for the design, construction and commissioning;
- failure to meet the required technical standards for quality, function, fitness for purpose, safety and environment preservation.

In most situations, the effect of adverse events will be financial loss. The task of the quantity surveyor is to identify the discrete sources of risk which cause failure to occur, and to develop a risk management strategy that indicates which organisation is best able to carry that risk.

Risk and uncertainty do not occur only on major projects. Whilst size is an important consideration, factors such as location, complexity, buildability, and type of building can all contribute to the risk. It is more likely that a complex, highly serviced five million pound operating theatre is going to carry more risk of a construction cost and time overrun than a five million pound warehouse building.

PROFESSIONAL ADVICE AND RISK

The consultant quantity surveyor acting for the client will always seek to balance the risks in the best interests of the client, whilst the contractor and specialist and trade contractors will also be seeking to balance their risk exposure. In general terms, the greater the risk a party must carry, the greater the reward that will be sought.

Consultants have an express and an implied professional duty of care to their clients, and in recent times clients have shown a willingness to sue for damages when they have suffered a loss as a result of bad advice. Where a consultant gives incorrect advice, he may be sued under the tort of negligence. A plaintiff must, of course, prove that the defendant owed him a duty of care, that this duty was broken and that damage was suffered as a result. The definition of skill and care is held to be 'such high standard as judged by practice and knowledge to be generally obtained at the date that the duties are performed'.

One argument often used is that consultants carry very little risk because they only give advice, but the loss that arises as a result of bad advice can be very significant. For instance, in the case of the high technology office buildings, losses for the client could be many orders of magnitude greater than the value of the work which might give rise to the breach or act of negligence.

RISK MANAGEMENT AND RISK ANALYSIS — THE PROCESS

Risk management evolved mainly in the USA to enable organisations to combat an ever increasing exposure to risk. Risk management is not synonymous with insurance, nor does it embrace the management of all risks to which a business is exposed. The main purpose of the tasks performed by the quantity surveyor must be to enable business, whether it be on behalf of the client or the contractor, to take the right risks. A simplified risk management system will involve the stages shown in Figure 5.1.

The purpose of this chapter is to discuss the techniques of risk analysis, hence there is no detailed discussion of the identification and classification of risk, nor on the response strategies. The quantity surveyor is dealing primarily with the cost, time, and quality aspects of a building project and in the following sections the discussion is concerned with risk analysis applied to the cost management of a project.

Cost is dictated in two ways: either the client has a maximum budget of £x to spend on a project, or he wants to know how much Xm^2 of floor area will cost for a particular type of building. The quantity surveyor will make a forecast based upon the information available at the time. Frequently, the budget price is a single price figure, but having recognised that construction work is beset with risk, is it reasonable to expect the construction price forecast to be a single point estimate? A more realistic approach is to adopt the policy that the single point forecast is the most likely price with a range between the lowest price and the highest price, given that certain conditions exist. The lowest price assumes that there are no unforeseen difficulties where the project has been completely designed prior to tender, whereas the highest price assumes the worst scenario for each event.

In practice, the consultant quantity surveyor will make a price forecast based upon certain known and some assumed factors about the design, and about the

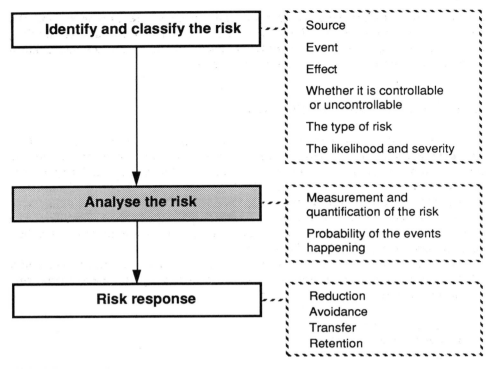

Figure 5.1 The risk management process

construction market. The factors will be both controllable and uncontrollable. For example, if the price of marble goes up then the design team can explore alternative options for the finishings; the situation is within the design team's control. However, if there is a requirement by the fire officer that certain stringent measures, which will substantially increase the cost of the work, must be incorporated into the building or a fire certificate will not be issued, then the design team have no option but to comply. In reality the situation is uncontrollable and outside the client and design team's control.

Risk analysis will not solve these difficulties but it will alert the client and the design team to the possibilities.

An argument against using risk analysis and giving a price range, is that the client who is not familiar with the construction industry could abandon the project at the early design stage if the highest price is above the price that he could possibly afford. Risk analysis is not a passive tool to warn of impending doom, it is intended to identify and quantify the areas where the design team might focus action to avoid or reduce the risks in the project and thus make the project a reality. It is much better that the client be made aware of likely cost variations at an early stage. Nothing creates a worse impression than for the client to be presented with extra costs during either the design or construction process.

The plea for using risk analysis in construction work is that the range and possible outcomes for each element or trade package are examined, and the cumulative effect on the overall cost is considered. Some risks, such as inflation, are beyond the control of the quantity surveyor, but that does not mean that a range of possible inflation rates should not be considered.

RISK ANALYSIS USING PROBABILITIES

Some risk cannot be confidently costed at the design stage, such as the effect on cost that exceptionally inclement weather will have on a project due to the foundation work commencing on site in December. Whereas the majority of risks arise from matters where there is a lack of information, for example, insufficient design and specification information at the early stages of design. As more information becomes available during the design phase, so many of the risks can be resolved until, just prior to a tender being sought, the estimate of construction cost contains only residual levels of risk.

A straightforward approach to including an allowance for the risk is to identify a list of risk items and assign each item with the probability of the event occurring and to give a three point estimate of:

- the most likely price;
- the lowest price;
- the highest price.

A typical risk item might be the probability of the need for a new gas main to be laid to the site. At the design stage the existing gas main had not been exposed hence the condition was not known. The most likely price might include an allowance for some modification to the existing main, whereas the lowest price could assume no work to the existing main is required, and the worst case that substantial work is needed to modify the main. Probabilities for each event would be assigned to each event such that there is 0.50 chance that some modifications are required, a 0.30 chance that no work is required and a 0.20 chance that substantial work will be needed.

The assessment of probability can rarely be an exact science, therefore expert judgement and intuition are required. In the above situation should the client be made aware of the worst likely eventuality or should an 'average risk allowance' be included in the budget forecast?

For example:

Item	Price	Probability	Allowance
A Some modifications to the existing gas main	£5000	0.50	£2500
B No modifications required other than inspection	£2000	0.30	£600
C Substantial modifications required to the gas main	£15 000	0.20	£3000
TOTAL			£6100

The quantity surveyor must consider what premium should be added to the budget price to ensure that the client has some comfort that all the eventualities have been taken into account. It might seem illogical to include £6100 which would cover both options A and B, but it must also be borne in mind that even the prices of £2000, £5000 and £15 000 will be informed best guesses based upon the information available.

There is a 0.20 probability that option C will be required. A more sophisticated approach uses a process called Monte Carlo simulation which is discussed after the overview of price prediction.

PRICE PREDICTION: AN OVERVIEW OF CURRENT PRACTICE

Construction prices used in forecasting are often based on the analysis of a small sample of historical projects for which the quantity surveyor has cost analyses or some form of cost breakdown, and which bear a close resemblance to the proposed building project. It is assumed that the tender price of an item to be built in the future can be determined by analysis and adjustment of the tender prices of analogous items built in the past.

Several factors interact and affect the reliability of the quantity surveyor's price forecast, including:

- the extent of design information available (ambiguity in the design and ambiguity in the prediction go hand in hand, whatever the method of prediction);
- the availability of reliable historical price data related to the type of project under consideration;
- the consultant quantity surveyor's familiarity with the type of project in hand and projects of a similar nature.

It is important to note that the prices used in the forecast can only be as good as the sample on which they are based. Further, the 'goodness' or otherwise of the sample is at least a two-dimensional concept relating to the size of the sample and the homogeneity of the sample. Other things being equal, it is always desirable that the sample should be as large as possible. On the other hand, it is important that the sample contains only construction projects that resemble the proposed project. In other words, the sample should be reasonably homogeneous with respect to the major cost significant features of the project.

Figure 5.2 outlines the information used in the cost planning process. Whilst we have concentrated on the use of historical cost information from completed projects, there are other sources of data that the quantity surveyor will use.

It has been suggested that a significant improvement in reliability can be obtained if historical price data are drawn from several buildings rather than from one, even if this means sacrificing some comparability; the exception to this is where there is a single identical completed building. Since the available database is finite, for instance, in house cost analyses of completed buildings, or the Building Cost Information Service (BCIS) cost analyses, a trade-off is imposed

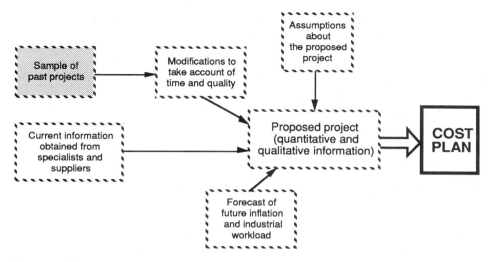

Figure 5.2 Outline of the cost planning process

between sample size and homogeneity. Unfortunately, the precise nature of this trade-off is unknown, but as the uncertainties are high, a limited number of samples (for instance, fewer than five) are probably inadequate.

A word of caution is worth noting here. Often, cost data relate to the analysis of the tender price whereas the information needed is the final account price of the completed project. Furthermore, cost data mask the impact of regional differences in construction prices and differences in the size, quality, complexity of design and construction and buildability of projects. The professional skill and judgement of the quantity surveyor is therefore needed in the careful selection of projects similar to the proposed project.

A further point to be considered is the interdependence of the elemental categories used in cost planning. Research has shown that certain of the elemental categories are not mutually exclusive. Some of the elements will be interdependent, for instance the cost of the electrical installation is likely to be higher in a building with air conditioning and where the mechanical services cost element is high. Any risk analysis programme does not take account of this interdependence between the elements and the only practical way this can be accommodated is by careful examination of the sample. Inevitably this is a criticism of any technique which uses a historical sample of price data for cost planning.

There are numerous different estimating techniques used at the design stage by quantity surveyors. We shall concentrate on one of these techniques, elemental cost planning using the BCIS elemental categories. The elemental unit quantities are measured for each selected elemental category and the historical sample is used to generate elemental unit price rates, which are based upon the arithmetic mean or some measure of central tendency. The derived unit price rates for each element are multiplied by the appropriate elemental quantity to build up the price prediction of the likely tender price.

In this prediction procedure, each calculated mean unit price rate represents a sample mean from a probability distribution. A different set of historical cost analyses or price data is likely to generate a different mean unit price rate, that is, a different sample mean from the same probability distribution. It is to be expected, of course, that there would be some offsetting changes in the mean unit price rates — some rising and others falling — but the resulting total price forecast represents only one possible value from a family of values.

No direct evidence can be obtained, at any time, of the way in which a prediction would be affected if it were based on a different, but equally homogeneous, sample; if such samples were available, there would be nothing to be gained by excluding them from the initial forecasting process. Indirect evidence of the probable variability of a prediction can be obtained, however, by using the variability of the calculated mean unit price rates. No matter how homogeneous the sample from which the elemental mean unit price rates are calculated, as long as it contains more than one building, the individual unit price rates will exhibit some residual variation about the 'true' mean unit price rates. It should be possible to use these residual variations to generate a picture of the probability distribution of the final prediction.

In other words, since each estimated unit price rate is drawn from a probability distribution, it follows that the overall forecast is also a member of a probability distribution, the characteristics of this distribution being determined by the characteristics of the individual distributions for each elemental category. Using the analysis outlined below, it is possible to approximate the probability distribution of the overall prediction to identify the characteristics of the family from which it is drawn. This in turn will allow us to identify:

● the probability that the contractor's tender price will not exceed the consultant quantity surveyor's price prediction;
● the most likely range within which the contractor's tender price will lie.

USING SIMULATION

The aim of this section is to describe Monte Carlo simulation and to use a technique which identifies the probability distribution from which a price prediction has been taken. The general methodology is applicable at any stage of the investment appraisal, development appraisal, or design process of a project. However, its use is perhaps best illustrated by applying it to the cost plan prepared at the scheme design stage by the quantity surveyor.

The single factor that characterises all price forecasting is uncertainty. Price prediction is not a precise scientific exercise, but an art which involves both intuition and expert judgement. Despite the undoubted desirability of an unbiased price prediction, there exists no objective test of the probability that a particular forecast will be achieved. Since a price prediction is the sum of many parts, any such objective evaluation of its precision is possible only by the use of statistical techniques.

Simulation is a word which is in common use today. The term simulation describes a wealth of varied and useful techniques, all connected with the mimicking of the rules of a model of some kind. Simulation techniques are used extensively in industry. For instance, flight simulators, used to train pilots, can introduce all types of hazards to the pilots by simulating the live situation. In construction, simulation techniques are used by the contractor at the tender stage to examine the impact of different weather patterns on the proposed construction programme.

Monte Carlo analysis is a form of stochastic simulation. Stochastic or probalistic means that the technique is concerned with controlling factors that cannot be estimated with certainty. It is called Monte Carlo because it makes use of random numbers to select outcomes, rather as a ball on a roulette wheel stops, theoretically at random, to select a winning number. Random numbers could be selected in a variety of ways, such as picking a number out of a hat, or throwing a die. In reality, using a computer program is the most effective method of generating sets of random numbers.

Probability theory allows future uncertainty to be expressed by a number, so that the uncertainty of different events may be directly compared. Information about the probability of a future event occurring, or a condition existing, is generally presented in the form of a probability density function. If then, we can obtain some indication of the probability density function to which a particular price prediction belongs, we have available a test of the likelihood that the estimate is unbiased.

RISK ANALYSIS USING MONTE CARLO SIMULATION

Risk analysis by means of Monte Carlo simulation can be defined as the study of the relationship between an estimated price and the chance or probability of the tender price deviating from that amount.

Monte Carlo simulation generates hypothetical mean unit price rates for each elemental category in the cost plan for the proposed building. These hypothetical rates are taken from probability distributions with the same statistical properties, that is, probability density functions, as those which characterise the original sample data from which the mean unit price rates were estimated. The hypothetical rates are then used to build up a total price forecast for the proposed building. If this exercise is repeated a sufficiently large number of times, it will be possible to obtain a picture of the probability density function which characterises the total price, and so to identify the most likely total price.

A STEP BY STEP APPROACH TO MONTE CARLO SIMULATION

The procedure is now described in more detail. Monte Carlo analysis proceeds by generating a series of simulations of a proposed project, each simulation giving a price prediction for the project. The predictions are plotted, first as a cumulative frequency curve and secondly as a histogram. There are several steps to the analysis.

Step 1

For any particular elemental category, such as the substructure, identify the probability distribution from which the price per square metre of the gross floor area used in the prediction is taken. We shall call this the *mean unit price rate* because the rate will be derived from several projects. This is the crucial and most difficult part of the analysis, particularly since we must make an a priori choice of the probability distribution. We have already shown why it is necessary to consider our sample of historical unit price rates as having been generated from some underlying probability distribution. The problem we now confront is that of choosing the appropriate probability distribution for a particular set of sample data. Our historical sample of unit price rates gives us measures of centrality (the mean) and dispersion (the variance) which the probability distribution must also exhibit. It is desirable, in addition, that the statistical distribution we use should include several other characteristics.

First, the distribution should be easily identifiable from a limited set of data, as a normal distribution is completely identified by mean and variance. This leads naturally to the second characteristic, that the distribution can be easily updated as additional historical data are introduced to the analysis.

Thirdly, the probability distribution should be flexible, that is, capable of taking on a wide variety of shapes. We might expect that a distribution of unit price rates will be skewed to the right as in Figure 5.3(a). In other words, we might expect extreme values for unit prices to be high rather than low, since the lower bound on the resource costs that underlie the calculation of unit price rate is rather more defined than the upper bound. A simple example will illustrate this point. If a vinyl tile floor finishing is considered at £15 per square metre, it is more likely to cost £30 per square metre than it is to cost nothing. Nevertheless, we should allow for situations in which there is no discernible skewness, or where the skewness is in the opposite direction, or where skewness is so strong as to generate a distribution such as that illustrated in Figure 5.3(b).

At the same time, there is no reason to believe that the appropriate probability distribution for one elemental category, for instance, substructure, needs to be the same as that for another elemental category, such as the external walls. It is desirable, therefore, that the probability distribution exhibits a richness in the variety of shapes it can adopt.

Finally, we would prefer that the probability distribution has finite end points which can be individually chosen. Moreover, we would expect that the economic decisions which unit price rates summarise — resources, design decisions, profit margins — are such as to impose lower and upper bounds on feasible or acceptable unit price rates for each elemental category. The probability distribution to be chosen should be capable of being defined within such bounds.

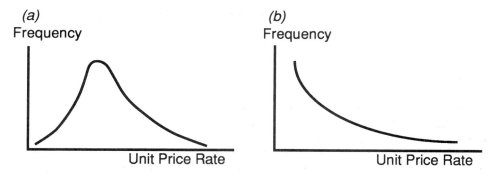

Figure 5.3 Skewed probability distribution

There are various distributions which can be used. The more simplistic approaches to Monte Carlo simulation use the uniform distribution, the triangular distribution, or the normal distribution as shown in Figure 5.4. However, the only set of probability distributions which adequately displays all of these characteristics is that set referred to as the beta distribution. Consider, for example, an alternative such as the normal distribution.

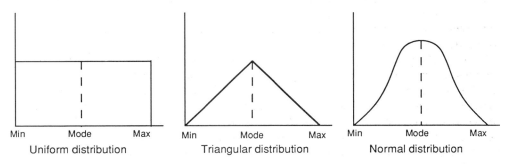

Figure 5.4 Types of distribution

The major limitations of the normal distribution are, first, that it has an infinite range. Secondly, and rather more importantly, the normal distribution is bell-shaped and so cannot accommodate skewness in the data.

Other distributions, such as Poisson, lognormal, binomial or uniform, have similar limitations. Rather than enumerate these, we will show that the beta distribution, on the other hand possesses the characteristics we noted above.

The beta distribution has the equation:

$$P(X) = \frac{1}{B\,(p,q)} \cdot \frac{(x-a)^{\,p-1}\,(b-x)^{\,q-1}}{(b-a)^{\,p+q-1}} \qquad (5.1)$$

$(a \le x \le b);\ p,q > 0$

where: $P(X)$ = frequency density function
         a = minimum price
         b = maximum price
         p, q = parameters of the distribution; $p, q > 0$
   $B(p,q)$ = beta function

The first point to note is that the appropriate beta distribution is determined completely by the parameters p, q, a and b, parameters which themselves are generated easily from the actual data to which the distribution refers. We might take a to be the lowest value and b to be the highest value in our sample. The values for p and q are then calculated from the equations:

$$p = \left(\frac{\mu_1 - a}{b - a}\right)^2 \cdot \left(1 - \frac{\mu_1 - a}{b - a}\right) \cdot \left(\frac{\mu_2}{(b - a)^2}\right)^{-1} - \left(\frac{\mu_1 - a}{b - a}\right) \tag{5.2}$$

$$q = \left(\frac{\mu_1 - a}{b - a}\right) \cdot \left(1 - \frac{\mu_1 - a}{b - a}\right)^2 \cdot \left(\frac{\mu_2}{(b - a)^2}\right)^{-1} - \left(1 - \frac{\mu_1 - a}{b - a}\right) \tag{5.3}$$

where: μ_1 = mean
       μ_2 = variance

It follows that the beta distribution for a particular elemental category can be quickly and easily updated if additional historical data are added to the sample.

It may be useful to illustrate the selection of a beta distribution by means of a hypothetical example. Assume that mean unit price rates for the electrical installation and mechanical installation elements have been generated from the analysis of ten completed buildings with unit price rates as given in Table 5.1.

Table 5.1 Hypothetical unit price rates

	1	2	3	4	5	6	7	8	9	10
Electrical installation	12.00	10.00	12.50	13.50	14.50	11.00	13.00	14.50	15.00	16.00
Mechanical installation	19.00	20.00	22.50	34.00	25.00	21.00	23.10	23.70	24.00	20.50

Electrical installation	mean unit price rate	£13.20
	variance	3.51
Mechanical installation	mean unit price rate	£23.28
	variance	17.92

These data can be used to identify the ranges within which the mean unit price rates fall. As noted above, we approximate these ranges as running from the lowest to the highest unit price rates in our data, that is, to be £10/m²−£16/m² for electrical installation and £19/m²−£34/m² for mechanical installation. Substituting these figures for mean, variance, maximum and minimum in equations (5.2) and (5.3), we estimate the beta distributions for electrical installation and mechanical installation as shown in Table 5.2. The values for p and q in Table 5.2 have been derived from the equations (5.2) and (5.3).

Table 5.2 Estimated beta distribution

Parameter	Electrical installation	Mechanical installation
a	£10	£19
b	£16	£34
p	0.83	0.45
q	0.72	1.11

Figure 5.5 shows in diagrammatic form the approach for step 1. As can be seen from the shape of the distributions for the various elements, they can, and do, take a variety of shapes.

Step 2

Having identified a beta distribution for each elemental category, generate a random number from each of these distributions. This is most easily achieved by using a pseudo random number generator on a computer. Each such random number is an estimate of the unit price rate for the appropriate elemental category. In our hypothetical example, the number generated for the electrical installation might be £12.75.

Note that this number need not equal any of the actual observations. It must, however, lie within the range (a, b) that is, £10m^2–£16m^2, and have an expected value equal to the mean unit price rate, that is, £13.20m^2.

Step 3

Multiply the random unit price rates by the measured quantities in the appropriate elemental categories for the proposed building, for example, if the measured quantity of the floor area for the electrical installation is 1000 m^2, we obtain £12.75 × 1000 = £12 750.

Step 4

Sum the results of step 3 for each of the elemental categories used in the analysis, to give a forecast of the total project price. Store this estimate and return to step 2. Repeat N times, where typically N = 50 100 200..., to generate N simulations of the project.

Step 5

Plot the N estimates as a cumulative frequency curve and as a histogram.

The choice of N is an immediate problem. It is preferable that N is 'large', since

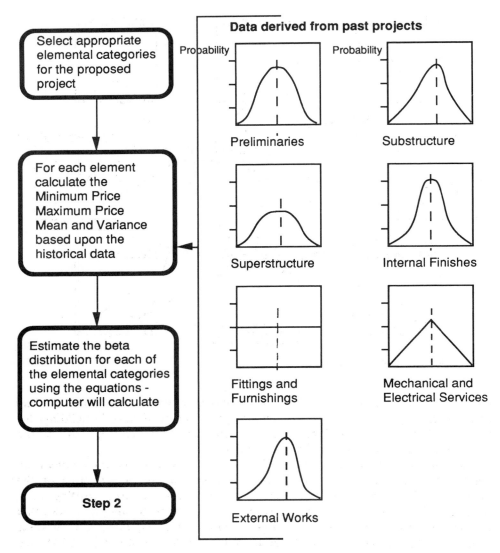

Figure 5.5 The approach for step 1

this will lead to a smoother cumulative frequency curve and histogram. On the other hand, additional simulations are not without cost, either in computer time or manual processing.

Figure 5.6 illustrates steps 2 to 5 in diagrammatic form.

USING MONTE CARLO SIMULATION ON A LIVE PROJECT

This process was undertaken for a live project and the results are presented in Figure 5.7. Each x on the diagram represents an estimate produced by the

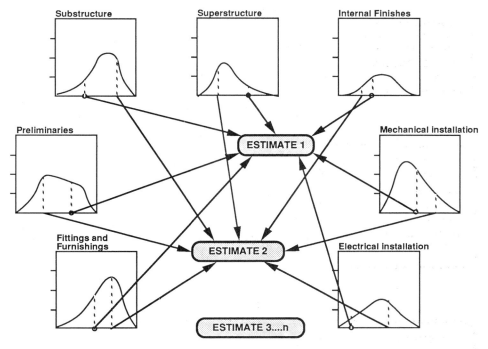

Figure 5.6 Steps 2 to 5 illustrated

computer. The proposed project was a warehouse building located in an urban area. Six homogeneous completed buildings (the number of suitable buildings for which cost data were available) were selected from the cost records of the quantity surveyor concerned with the building.

The cumulative frequency curve in Figure 5.7 shows that 500 simulations were conducted. This curve is used to derive the probability of the price of the proposed building falling within a specified range. For example, reading along from the 250 point on the vertical axis there is a 50% probability (250/500) that the building price will be less than approximately £481 000, while reading along from the 400 point on the vertical axis, there is an 80% probability (400/500) that the building price will be less than £531 000.

The histogram in Figure 5.8 is used to supplement the cumulative frequency curve, since it indicates the 'most likely' price range for the proposed building. In this case, the most likely building price is between £470 000 and £496 000. When the simulation process has been completed and a cumulative frequency curve and histogram generated, these can be used to assess the single price forecast produced by the quantity surveyor using the unit price rates for the historical sample of buildings (see above).

The single price prediction produced by the consultant quantity surveyor for the project was £461 000. The cumulative frequency curve shows that there is a 30% probability that the tender price will not exceed the price forecast, while the histogram indicates that there is approximately a 60% probability that the tender

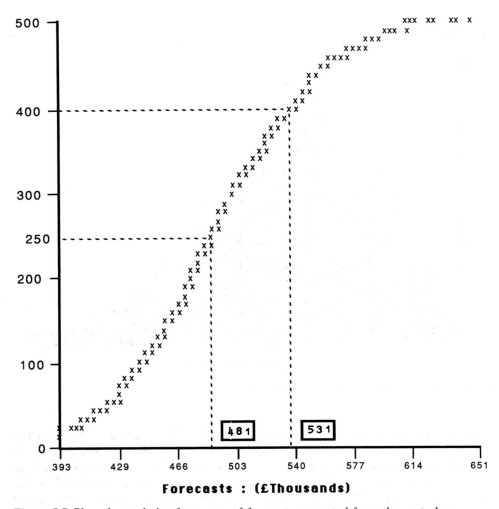

Figure 5.7 Plot of cumulative frequency of forecasts generated from the cost plan

will be within 10 % of this prediction. There is no objective way of stating whether these percentages are good or bad, although it would appear that the quantity surveyor's forecast of £461 000 was slightly low. In this sense, risk analysis is by no means a substitute for the personal judgement of the quantity surveyor. However, it is a method that identifies situations in which single price forecasts should be subjected to close scrutiny.

High variance in the cumulative frequency curve and histogram results from high variance in the original data. To see why this is so, it should be noted that since the total price forecasts as shown in Figure 5.7 and 5.8 are generated from the summation of a series of distributions, that is, one distribution for each elemental category, the total price forecasts themselves are distributed as a beta

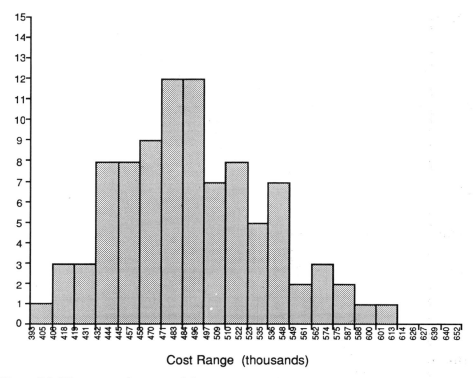

Figure 5.8 Histogram of generated forecasts

distribution. It follows that the predictions will exhibit the same skew, to the right, as those of the individual elemental categories, and a variance which is related to the variances of the mean unit price rates for these categories. This implies that the more variable the historical data, the more dispersed will be the generated estimates. Hence the slope of the cumulative frequency curve will be flatter and wider the most likely the price range becomes.

The result

The project was bid by five contractors. The bids received were as follows: £472 500, £496 000, £502 000, £507 000, £551 000. The lowest tender included £102 000 of prime cost and provisional sums in the bill of quantities.

CONCLUSIONS

There are many commercial software packages available on the market that will undertake risk analysis using Monte Carlo simulation. Risk analysis does not provide the solution to eliminating risk in the forecasting of construction price it is a valuable tool that encourages the design team to focus on the 'what if' questions and to consider the range of possibilities that might occur.

In order for risk analysis to be adopted by quantity surveyors it must prove itself to be economically viable for both clients and the professional advisers. Risk analysis must be seen as a practical tool that enhances existing skills. The technique is widely used in other industries; given the opportunity it will become an everyday technique within the construction industry.

BIBLIOGRAPHY

Byrne, P. and Cadman, D. (1984) *Risk Uncertainty and Decision Making in Property Development*. London: E. & F.N. Spon Ltd.

Carter, E. E. (1972) What is risk in risk analysis. *Harvard Business Review*, July/August.

Chapman, C. B. and Cooper, D. F. (1983) Risk engineering: basic controlled interval and memory models. *Journal of Operational Research Society*, **341**, pp. 51–60.

Flanagan, R. and Norman, G. (1982) Risk Analysis — an extension of price prediction techniques for building work. *Construction Papers*, **1**, 3, pp. 27–36.

Hertz, D. B. and Thomas, H. (1983) *Risk Analysis and its Applications*. New York: John Wiley and Sons.

Hertz, D. B. (1979) Risk analysis in capital investment. *Harvard Business Review*, September/October, pp. 169–182.

Jardine Insurance (1984) *Risk Management: Practical Techniques to Minimise Exposure to Accidental Losses*. London: Kogan Page Ltd/Chartered Institute of Management Accountants.

Johnson, N. L. and Kotz, K. (1970) *Continuous Univariate Distributions*, **2**. USA: Houghton-Mifflin Company, pp. 37–51.

Singh, G. H. and Kiangi, G. (1987) *Risk and Reliability Appraisal on Microcomputers*. Bromley: Chartwell-Bratt Ltd.

Woods, D. H. (1966) Improving estimates that involve uncertainty. *Harvard Business Review*, 44, pp. 91–98.

Chapter 6

Life Cycle Cost Management

DAVID HOAR, *Directing Surveyor, Nottinghamshire County Council* and GEORGE NORMAN, *Tyler Professor of Economics, University of Leicester*

INTRODUCTION AND DESCRIPTION OF TECHNIQUES

Brief history of development of techniques

During the last two decades there has been a growing awareness in the United Kingdom of the importance of considering the costs of buildings in use and of developing financial techniques to evaluate the whole life costs in buildings and components in use.

In the late 1970s life cycle costing was actively encouraged by the United States federal government, because they were becoming increasingly concerned about the costs in use of large federal investments across the United States. Prior to that time, central government had been concerned solely with controlling the capital inputs into federally financed projects. However, the central government of the United States were finding that they had little control of those investments once the buildings had been commissioned. The United Stages, a large country with very differing climatic conditions, poses special problems for a central government in the allocation of resources. It was for this reason that in the late 1970s the federal government insisted that any federal investments in building projects would only be made following the submission of a life cycle costing exercise that would assess the cost of the buildings in use.

In 1981, the Quantity Surveyors Division of the Royal Institution of Chartered Surveyors commissioned the Department of Construction Management of the University of Reading to undertake a study into life cycle costing for construction. In 1983 Roger Flanagan and George Norman, both at that time at the University of Reading, produced their publication *Life Cycle Costing for Construction*.

Shortly after the publication of *Life Cycle Costing for Construction* the Quantity Surveyors Division of the Royal Institution of Chartered Surveyors became concerned that the technique was not being adopted by the profession. They established a working party to endeavour to produce a practical working guide to life cycle costing techniques and a worked example to help practitioners to become acquainted with the techniques.

In 1986 the working party produced *A Guide to Life Cycle Costing for Construction* and followed this in 1987 with the publication entitled *Life Cycle Costing: a Worked Example*.

The technique is now becoming widely used particularly in the public sector where a number of government departments are insisting on the adoption of appraisals of project options prior to giving approval to projects. Life cycle costing techniques form the basis for these investment appraisals.

Definition

The life cycle cost of an asset is defined as the present value of the total cost of that asset over its operating life including initial capital costs, occupation costs, operating costs and the cost or benefit of the eventual disposal of the asset at the end of its life. The life cycle cost approach is therefore concerned with the time-stream of costs and benefits that flow throughout the life of a project. All future costs and benefits are converted to present values by the use of discounting techniques, and in this way the economic worth of an option can be assessed.

Life cycle cost planning, management and analysis

Life cycle cost planning is the identification of total costs of a building or an individual building element, which takes explicit account of initial capital costs, subsequent running costs and residual values, if appropriate, and expresses these various costs in a consistent and comparable manner by applying discounting techniques.

List cycle cost analysis is related to the technique of cost analysis practised by quantity surveyors in relation to capital projects. The technique is concerned with analysing the life cycle costs of a completed project in use after a period of time. Life cycle cost analysis is therefore a tool for quantity surveyors to adopt in assisting them to prepare life cycle cost plans for construction projects.

Life cycle cost management is derived from life cycle cost analysis and identifies those areas in which costs in use detailed by life cycle cost analysis may be reduced. Life cycle cost management can therefore be used to assist building owners in comparing buildings. The following examples are given to illustrate the types of benefits that can be derived from the use of life cycle cost management techniques:

- To identify the efficient or inefficient utilisation of buildings.
- To identify the energy efficiency of buildings.
- To identify the differences in maintenance costs between similar buildings and to assist in building maintenance management.
- To provide taxation advice to clients on items related to buildings in use.

The concept of levels for life cycle costs

Life Cycle Costing for Construction recognised that the technique of life cycle costing involves the manipulation of a large amount of data. This data can be grouped into a hierarchical structure where items are classified to successive sub-divisions by consistent reference to a single common attribute at each stage.

Figures 6.1 and 6.2 illustrate the concept of levels. These diagrams should be self explanatory. The diagrams also show the types of life cycle costing activities that can be undertaken at each of the levels.

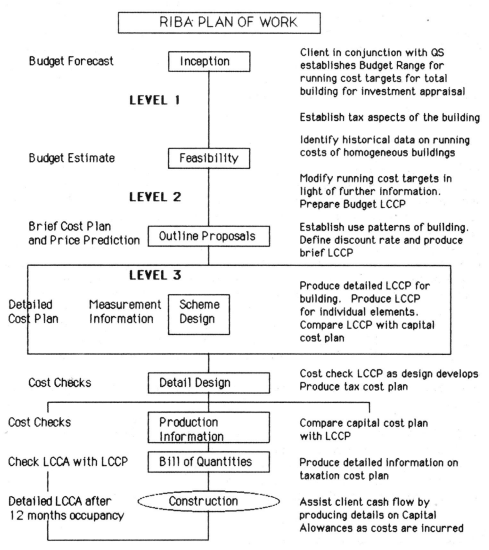

Figure 6.1 Life cycle cost and the RIBA plan of work. Source: Flanagan and Norman (1983)

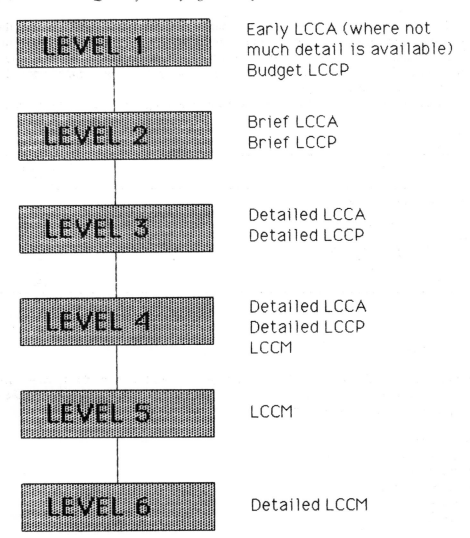

Figure 6.2 The relationship of levels. Source: Flanagan and Norman (1983)

Life of the investment

An essential element in the life cycle costing appraisal is to determine the life of the investment. Typically the life of the investment is the period over which the organisation for whom the building is being commissioned is expected to hold an interest in the building. At the end of the life of the investment the building and the land will have a residual value.

Quantity surveyors engaged in life cycle cost planning will need to consult their clients and their client's other professional advisers such as accountants, in relation to the 'life of the investment' for any particular commission. The life of the investment will largely be determined by the nature of the client organisation, the type of project being considered and the risk associated with that project. Project lives usually lie within the range three years to sixty years, dependent upon the nature of the project.

Typical project lives used for life cycle cost planning are given in Table 6.1.

Table 6.1

Project	Life of investment (years)
Fast food high street outlet by international market leader	3
Factory for car manufacturer	10
Office development undertaken by property development company	10
School for local authority	20
Prison for Home Office	60
Cathedral	100

Obsolescence

Building life is influenced by obsolescence. Almost all forms of obsolescence are related to economic considerations. *Life Cycle Costing for Construction* identified six different forms of obsolescence:

- physical
- economic
- functional
- technological
- social
- legal.

Physical obsolescence is reached when a building is likely to collapse due to structural failure.

Economic obsolescence is achieved when occupation of the building is not considered to be the least cost alternative of meeting a particular objective. A good example of a building and related site reaching the end of its economic life would be a football ground in a high value city centre location that would be suitable for redevelopment for quality housing or upmarket retail outlets.

The functional life of a building ends when it ceases to function for the same purpose as that for which it was built. An example of this type of change is the use of a cinema that has been converted into a snooker hall.

The technological life of a building ends when it is no longer technologically superior to alternatives. An example would be where a high tech computing or electronic company for prestige and operational reasons needs an office that accommodates advanced technology. When a building can no longer accommodate high tech equipment due to its physical constraints it reaches the end of its technological life.

The social or legal life ends when popular demand or legal requirements dictate a replacement for reasons other than economic considerations. An example may be the use of timber football stands following the Bradford football stand fire disaster.

The life of an investment for a particular life cycle cost study should be given careful consideration by the quantity surveyor in consultation with the client. The life of the investment selected will therefore usually be determined by the client with the advice of the quantity surveyor and his other professional advisers.

Discounting

Life cycle costing takes explicit account of the fact that investment in durable assets such as buildings and building components commits the owner/operator to both current and future cost expenditures. Any method for appraising different design or management options must take both sets of costs into account and should compare them on a common basis. What this implies is that any acceptable investment appraisal technique must exhibit three properties:

(1) it should take account of *all* cash flows associated with the investment throughout the life of the investment;
(2) it must make proper allowance for the time value of money;
(3) the (notional) return on the investment should be not less than the market rate of interest.

Life cycle costing by definition satisfies the first of these criteria. This section concentrates upon the second criterion and we shall consider the third in a later section. The first point we should consider is just what we mean by the time value of money. Very simply, money has time value because the worth to us of a sum of money is dependent upon when that sum either becomes available or has to be expended. To see why this is so consider the following question:

'I need to spend £1050 (at today's prices) next year on the installation of a partial replacement to the heating system. I am preparing my budget to-day. What sum should be set aside today in order to meet the anticipated expenditure?'

In order to answer this question it is necessary to know at what interest rate current money can be invested. Assume that the net of tax and net of inflation interest rate is 5%. Then if I commit £1000 today this will be sufficient to meet the anticipated expenditure. Put another way, with a real interest rate of 5% an expenditure of £1050 in one year's time is equivalent to an expenditure of £1000 today. The interest rate is effectively being used as an *exchange rate* between current and future money.

This can be expressed in more general terms. With an interest rate of r% an expenditure of £C today is equivalent to an expenditure of £C $(1 + r)$ in one year's time. Or, an expenditure of £T_1 in one year's time is equivalent to a *present* expenditure of:

$$C_1 = T_1/(1 + r)$$

Now consider the situation in which the heating system is forecast to need regular maintenance with expenditures at today's prices of £200 per annum. We again will want to know what sum of money has to be committed today in order to meet these forecast expenditures. Assume once more that the interest rate is 5%. The argument above indicates that the present value of the maintenance expenditures next year is $C_1 = T_1/(1 + r)$, i.e. $C = 200/(1.05) = £190.48$. Now apply the same argument to the expenditures in year 2. This means that the £200 to be expended in year 2 is equivalent to an expenditure of £190.48 in year 1. Now we apply the present value conversion to this sum: £190.48 in year 1 has present value $190.48/(1.05) = £181.41$. In other words, the £200 to be expended in year 2 has present value $200/(1.05 \times 1.05) = 200/(1.05)^2 = £181.41$. More generally, the present value of an expenditure of T_2 in year 2 at interest rate r% is:

$$C_2 = T_2/(1 + r)^2$$

This discussion can be extended to expenditures incurred at any point in the future. The present value of a sum T_t expended in year t with market interest rate r% is:

$$C_t = T_t/(1 + r)^t$$

and the total present value of the maintenance expenditures over the life cycle (N years say) of the heating system is:

$$PV = C_0 + C_1 + C_2 + C_3 + C_4 \ldots + C_t + \ldots C_N$$
$$= T_0 + T_1/(1 + r) + T_2/(1 + r)^2 + \ldots + T_t/(1 + r)^t + \ldots T_N/(1 + r)^N$$

The procedure we have just demonstrated is called discounting and the interest rate is referred to as the discount rate.

One of the main obstacles to the implementation of life cycle costing has undoubtedly been suspicion about the meaning of discounted money and present values. Some consider the procedure above to be a complicated piece of arithmetic with little real meaning. Nothing, however, could be further from the truth. Indeed, we can give the concept of present value a precise economic interpretation. As we have tried to point out in the calculations, it represents the sum of money which, if put aside today, would just cover all forecast expenditure on a particular building or building component over the projected life cycle of that building or component. What this means is that the present value of these forecast expenditures is a price. It is the present exchange value, or price of the forecast expenditures given the interest rate, or discount rate, at which they are calculated. And as we shall see below, the discount rate should generally be the market rate of interest that the decision-maker knows has to be paid.

A further implication of this discussion is that the present value of a future sum is less the further away in time the sum is due to be received or expended.

What all this amounts to is the idea that the value of money is time dependent, dependent upon the precise timing of the expenditures or receipts. It also means that we now have a technique that will allow us to compare options with different time-streams of expenditures and receipts. It is incorrect to calculate the total cost of an option — whether this be a design option for a complete building, or a proposed replacement energy management system, for example, merely by adding up all of the costs net of any receipts. To do so would be to ignore the timing of these expenditures and receipts. Rather, the expenditures and receipts should first be converted to present values and then added up. The price implicit in the time value of money and the discounting process effect this conversion.

Risk and uncertainty

We have seen that life cycle costing deals with the future. But the future is, by definition, unknown. Cost estimates for maintenance, refurbishment, cleaning and so on are precisely that — estimates. They cannot be known with certainty no matter how good the database from which they have been forecast. This uncertainty in the cost inputs implies that the output of life cycle costing — a forecast life cycle cost of an option — is also uncertain. This has caused some organisations to reject life cycle costing on the basis that the results appear accurate given the mathematics involved in their calculation, but are inherently guess work.

This argument is wrong as it stands, but does have a sufficient grain of truth to require some response. It is here that risk management has an essential role to play. We do not propose to go into this in any detail in this chapter. The reader is referred to Chapter 5 in which a rather more detailed description of these techniques is to be found. Some brief comments can be made at this point. The essence of risk analysis and risk management is to allow the decision-maker to use the risk and uncertainty in order to improve decision-making. This approach has two elements to it: sensitivity analysis and probability analysis.

The distinction between these two elements is relatively straightforward. Sensitivity analysis does not, in general, require that a probability distribution be associated with each estimate that is subject to risk. Rather it is a deterministic technique that asks repeated 'what if' questions. Probability analysis, of which the dominant technique is Monte Carlo simulation, by contrast treats uncertainty in an explicit and multi-dimensional manner. All variables subject to risk are modelled as probability distributions rather than as known single values.

The output of a sensitivity analysis shows the extent to which the forecast life cycle cost of an option is sensitive to change in one of the forecast elements of the life cycle costing exercise: this may be, for example, the discount rate, the project life, or any one of the forecast cost streams. Since the major use of life cycle costing is to rank the options according to a well defined criterion — lowest life cycle cost — sensitivity analysis is of considerable use in determining whether the initial ranking is sensitive to changes in any of the forecast elements. We can use this technique, for example, to ask whether the choice of one energy management system in preference to another is sensitive to forecast discount rates, or project lives, or time-streams of expenditures. In many cases the initial rankings will be unaffected by even quite sharp changes in parameter values — allowing the decision-maker to place much more confidence in the recommended option. But there will be occasions when the ranking will change in response to a relatively small change in a parameter value. For example, consider two options that have very different patterns of costs over time, with the first perhaps having major cost expenditures after year 5 and the second having the major cost expenditures very early in its life. The ranking of these options is likely to be sensitive to the choice of discount rate. By contrast, the more similar the pattern of expenditures and receipts, the less sensitive will ranking be to the discount rate.

Probability analysis is, as has been indicated, multi-dimensional. It takes account of the uncertainty in all forecast elements and generates an overall view of the level of risk associated with a particular design option. This will typically take the form of a probability distribution of the total costs of the option which shows the most likely cost of the option and the range within which it can be expected to lie. The decision-maker will be able to identify from this distribution the probability that cost will lie within a defined range, or will not exceed a stated sum. Surprises are unlikely to be completely eliminated by the use of probability analysis, but they will certainly be considerably reduced.

OBJECT AND METHOD

Preselection of options

Life cycle cost planning analysis and management enables different options to be compared. The types of option appraisal that can be undertaken are summarised as:

- Comparison between alternative site and/or design proposals for new projects.
- Comparison between existing buildings in use to meet client's needs and proposals meeting these needs by way of new projects.
- Comparison between similar buildings of the client.
- Comparison between alternative building components or elements in new building projects.

In considering the generation of options to be evaluated in cases where the client's existing property portfolios will form part of that appraisal, it is a useful technique as a benchmark for further studies to evaluate the net present value of preserving the status quo in terms of buildings. As an example, if a client is occupying an office to administer his business and he is proposing to dispose of that office and establish his business in new office accommodation, it is fundamental to any investment appraisal that the client is aware of the costs in net present value terms of continuing to operate from his existing accommodation. The 'status quo option' should therefore always be assessed in cases where clients are operating from existing accommodation and are proposing to move into new or alternative accommodation.

The costs and values to be considered

The costs to be considered in a life cycle cost plan must include all items of expenditure required throughout the life of the building. The major categories of costs are:

- Capital costs
- Financing costs
- Operation costs
- Annual maintenance costs
- Intermittent maintenance, replacement and alterations costs
- Occupancy costs
- Residual values and disposal costs

The expanded check lists of costs given in *A Guide to Life Cycle Costing for Construction* is given in Appendix A. It is suggested that it is easier to deal with future costs if they are assessed at current prices. The level of detail used in assessing costs and values will largely be dependent upon the availability of information and the requirements of the client.

Particular care should be taken by the quantity surveyor in considering residual values and disposal costs. In instances where the residual values are likely to be significant in terms of value or because a relatively short period has been used for the life cycle cost appraisal, residual values should be assessed by a chartered valuation surveyor or other appropriate professional adviser.

Taxation allowances and incentives

This section is relevant to the effect of taxation allowances and incentives in the United Kingdom. Currently tax relief is available on both capital and revenue expenditure. Capital costs obtain tax relief by way of capital allowances. These allowances are deductible from the client's taxation account. Maintenance costs are a charge on the profit and loss account, which reduce the amount of tax payable.

The significance of tax relief on a particular project will depend upon the volume of the allowable expenditure. The impact of tax relief may have a very significant effect on the economics of any particular option that is being considered by the client.

It is not considered appropriate in this chapter to deal in detail with the matter of taxation allowances and incentives as they are subject to change by amendment to the Finance Acts. However, quantity surveyors undertaking life cycle cost planning for clients who are liable to taxation must ensure that they are adopting the appropriate capital allowances, as they may have a significant effect on the net present value of a particular option.

The discount rate

In our discussion of discounting reference was made to costs at today's prices and to an inflation-free and tax-free discount rate. We do not propose to deal with taxation allowances in any detail. Suffice it to say that if all cost and revenue estimates are made net of tax then the discount rate to be applied to these cash flows should also be net of tax. Inflation and its impact deserves more attention. There is a very close connection between the discount rate and inflation. Very simply, it is important to distinguish between the *interest rate* and the *discount rate*. The interest rate incorporates both the time value of money and the impact of inflation whereas the discount rate will typically abstract from inflation. The following equation shows this distinction:

$$r = \left[\frac{(1+d)}{(1+i)} - 1 \right] \times 100$$

where r = net of inflation discount rate (the real discount rate);
       d = interest rate including inflation (the nominal discount rate);
       i = inflation rate.

It is vital that for every option being analysed, cash flows be calculated on the same basis. For example, if cash flows are estimated in nominal terms, i.e. include an estimate for inflation, they should be discounted at a nominal discount rate, and this should apply to all of the options under consideration. On the other hand, if all estimates are in current prices excluding any allowance for inflation they should be discounted at the real discount rate.

It is difficult to be definitive regarding which approach is to be preferred. One simple rule can, however, be stated. If all cost estimates are expected to inflate at the same rate then it is probably preferable to perform all calculations in current prices applying a real discount rate. On the other hand, if inflation is expected to operate differentially across cost streams, e.g. if labour costs are expected to inflate at a very different rate from fuel costs or materials costs, then the calculations should be done in nominal terms with explicit account being taken of the differential rates of inflation.

No matter whether we are performing the discounting calculations in nominal or real terms, the appropriate choice of discount rate will depend upon the type of client, his objectives and the financial constraints he faces. The clearest distinctions to be drawn are between public and private sector clients, and between financing through borrowed money or from retained earnings.

In the case of financing through borrowing the appropriate discount rate is the actual cost of borrowing, net, of course, of inflation if the calculations are in real terms. By contrast, if the project is to be financed from capital assets, e.g. from retained income, or the sale of debt or equity, then the discount rate will be determined by the current and future rate of return that the market expects from that client's particular industry and firm. Ultimately in this case the discount rate is the opportunity cost of capital: the return to be expected from the best alternative use of the funds that are being committed to the particular project.

It will often be the case that the discount rate will be a critical variable in choosing between competing solutions to a particular problem, especially, as we noted earlier, when the various options being considered have very different patterns of cash flow. Too low a discount rate will bias decisions in favour of short-term low-capital-cost options, while too high a discount rate will give an undue bias to future cost savings at the expense of higher initial outlays.

As we have indicated, the choice of discount rate should reflect the particular circumstances. Nevertheless, we also recognise that at least some guidance should be given as to approximately what is an appropriate discount rate. In trying to answer this question, advantage can be taken of a roughly stable relationship between market interest rates and the inflation rate. This implies that a real discount rate in the region of 4−5% is probably correct if there is no better estimate available. Certainly, a real discount rate of 10% that has been applied by some private sector organisations is, in our view, far too high and cannot even be justified by appealing to the idea that the particular projects to which this discount rate is being applied are risky. In our opinion it is not appropriate to deal with risk by including a risk premium in the discount rate. Doing so carries the implication that risk cumulates over time, an implication that is difficult to sustain.

Presentation of results

The examination of a range of options using life cycle costing techniques can generate a large amount of detailed working. Quantity surveyors must ensure that in presenting reports to clients the clients are given the correct level of detail upon which they will make a management decision.

The level of detail given to a particular client will depend upon the type of client involved and his specific requirements. In instances where clients retain accountants and other professionals they may require detailed working papers to be furnished to support any presentation of results.

The main guideline for the quantity surveyor in presenting life cycle cost appraisals is to ensure that the presentation is given at the right level in terms of format and level of detail. A useful technique is to state the broad conclusions that have been established from an examination of options in summary form at the commencement of the report. The broad conclusions would then be followed by some supporting detail showing the summary of costs for each life cycle cost appraisal undertaken.

A worked example of a life cycle costing exercise is included later in this chapter, under maintenance management, for guidance purposes.

APPLICATIONS

Design choice

The team responsible for the design of a new construction project can significantly affect costs throughout the life of the project by the choice of the design. The strategic and tactical design decisions will have an impact on whole life costs. The design choice will affect all of the costs and categories listed in Appendix 1.

In order that the design team are aware of the life cycle cost implications of any particular design, a life cycle cost plan should be prepared for each of the major design options. In this way the design team will be able to appraise the building owner of:

(1) the capital cost of the project;
(2) the whole life cost of the project in terms of net present cost (based on a selected 'building life' for the project and an agreed discount rate).

Furthermore, life cycle cost plans can be used to produce cash flow forecasts for the first say three to five years, for each of the options to be considered.

Life cycle costing techniques should therefore be used as an integral part of the design process so that building owners are aware of the likely cost of adopting a particular design option in terms of capital cost, whole life cost and short term cash flow forecasts.

Maintenance management

Building maintenance represents a significant item of cost throughout the life of a building project. The cost significance of building maintenance will depend on a number of factors such as the construction type, the provision of mechanical and electrical services, and the function of the building. In relatively simple buildings, maintenance can typically represent 12.5 % of the total cost of a project throughout its building life.

Building maintenance managers are in a fundamentally different position to a building designer, in that they have to operate within the parameters set by the building owner in terms of disturbance caused to the occupier in fulfilling the building maintenance function. Furthermore, building maintenance managers must maintain the building irrespective of the difficulties, or otherwise, that may be imposed by the design and detail of the building.

Building maintenance managers will often have an element of choice in maintaining a building. Life cycle costing techniques can be adopted so that choices between alternative maintenance proposals for a particular structural element of the building can be made in an informed way, and having regard to whole life costs.

Although the building maintenance manager will primarily be concerned with making the correct choice in terms of building maintenance, he will also need to have regard for the implications that his maintenance decisions will have on other costs in use such as energy costs, cleaning costs and the implications of taxation allowances and incentives. The building maintenance manager will make his maintenance decisions against the background of the building type and function and the 'building life' required by the building owner.

As a simple example of the use of life cycle costing techniques that could be used by a building maintenance manager, the example of a bituminous felt flat roof covering is given.

Example

The building maintenance manager employed by a building owner with a significant property portfolio of buildings having flat roofs covered with bituminous felt roofing needs to formulate a policy for the replacement of those felt roofs. The building owner is seeking a building life for all of his building of 25 years. The building maintenance manager can consider a number of specifications but decides to restrict his choice to the following alternative specifications:

(1) Low cost specification
    Bitumen felt roofing
    Felt; first and second layers BS 747 bitumen fibre based type 1B weighing 18 kg/10 m^2; third layer bitumen fibre based type 1B weighing 25 kg/10 m^2; hot bitumen bonding compound; 50 mm laps; covering with 12 mm layer of 9 mm white limestone chippings in hot bitumen
    Covering; overall bonding first layer to insulation board base
    Over 300 mm wide; flat

(2) High performance specification
    Bitumen felt roofing
    Felt; first and second layers BS 747 glass fibre base type 3B fine sand surface finished weighing 18/kg 10 m^2; third layer Anderson Roofing HT350 sand finished weighing 34 kg/8.5 m^2; hot bitumen bonding compound; 50 mm laps; 30 mm ICI 'Purlroofer' insulation board underlay; bitumen felt vapour barrier BS 747 glass fibre based type 1B weighing 17 kg/10 m^2; overall bonding between layers; covering with 12 mm layer of 9 mm white limestone chippings in hot bitumen
    Covering; overall bonding first layer to insulation board base
    Over 300 mm wide; flat

Based on the adoption of a discount rate of 5% and the building maintenance manager's experience that low cost felt roofs give a component life of 10 years and high performance felt roofs give a component life of 20 years, the net present cost of these two options is calculated as shown in Table 6.2.

Table 6.2 Net present cost (roofs)

Item	£ cost per m² (not discounted)	Year	Discounting factor	Net present cost per m² (£)
Option 1 Low cost specification				
Initial cost	15.00	0	1	15.00
Replacement	17.00	10	0.6139	10.44
Replacement	17.00	20	0.3769	6.41
Total net present cost per m²				31.85
Option 2 High performance specification				
Initial cost	20.00	0	1	20.00
Replacement	22.00	20	0.3769	8.20
Total net present cost per m²				28.29

The example shows that based on the assumptions in relation to the discount rate, life of the investment and life of differing roofing materials, the preferred option is the high performance felt roofing specification, which has a lower net present cost of £28.29 compared with the low cost felt roofing specification at £31.85 per m² of roofing area.

Energy management

Those responsible for the energy management of buildings must, in the same way as building maintenance managers, work within the broad parameters of the building imposed by the design team.

Throughout the life of the building, the energy manager can consider proposals for improving the energy performance of buildings by introducing energy conservation measures.

The most cost effective energy conservation measures are likely to be those measures that require little or no capital input. Measures such as improved housekeeping and energy efficiency awareness by building owners will tend to be the most cost effective because they require little or no expenditure to achieve energy savings.

Typical energy saving profiles by introducing energy saving measures of different types are given in Figures 6.3–6.6 for illustrative purposes.

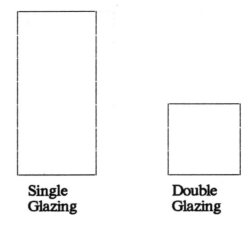

50% saving in heat lost through windows

Figure 6.3 Install double glazing

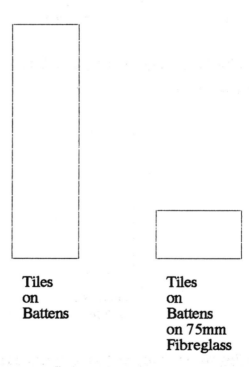

80% saving in heat lost through roof

Figure 6.4 Insulate tile and batten pitch roof with 75 mm fibreglass

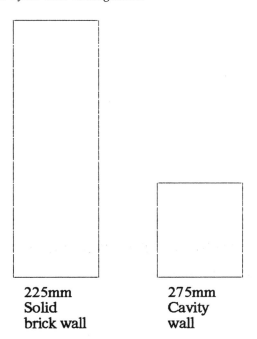

225mm
Solid
brick wall

275mm
Cavity
wall

<u>60% Saving in heat lost through walls</u>

Figure 6.5 Cavity walls

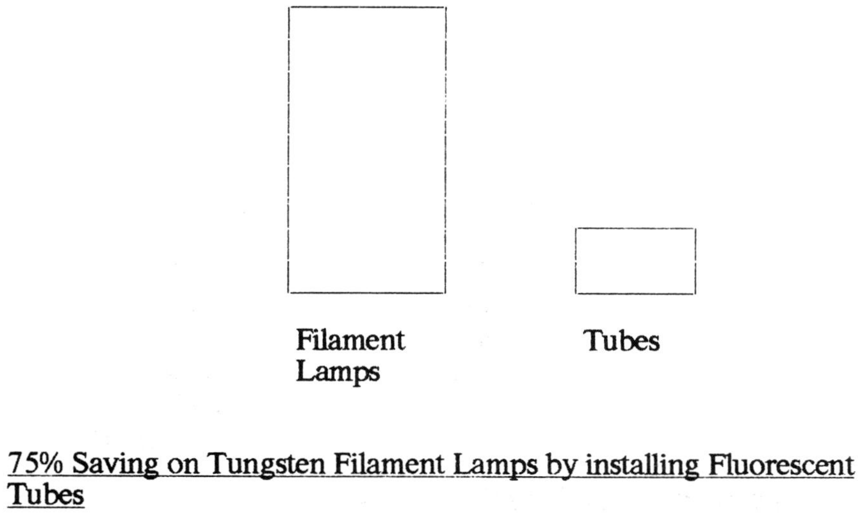

Filament
Lamps

Tubes

<u>75% Saving on Tungsten Filament Lamps by installing Fluorescent
Tubes</u>

Figure 6.6 Lighting

Cleaning

Those involved in managing the cleaning of buildings, like building maintenance managers and energy managers, have to clean the building that is presented to them by the design team and the buildings owner. Cleaning managers seldom have an opportunity to change the elements of the building in such a way as to reduce cleaning costs.

However, there are opportunities throughout the life of the building project when decisions can be made that will reduce the costs of cleaning. As an example, consider a large local authority office that had to replace floors covered with sheet vinyl flooring.

Example
The building maintenance manager and the cleaning manager employed by a local authority had to consider the replacement of floor finishes in the authority's principal administrative building. The building required a component life of 10 years.

Choice of floor finishes was restricted to the following specifications:

(1) Vinyl sheet
    Flexible sheet finishings
    Vinyl sheet; heavy duty; Marley 'HD' or similar; level; with welded seams; fixing with adhesive; to cement and sand base
    2 mm thick; over 300 mm wide

(2) Carpet tiles
    Heavy duty 'Burmatex Velour' carpet tiles; colour ref 241 Celtic Green; fixing with spot adhesive at four corners of each tile; all in accordance with manufacturers instructions
    500 mm × 500 mm × 8 mm; over 300 mm wide

Based on the adoption of a discount rate of 5%, the building maintenance manager's experience that both products would have a life of 10 years, and the cleaning manager's experience of the cost of wet cleaning sheet vinyl floors and vacuum cleaning carpet tiles, the net present cost of these two options is as shown in Table 6.3.

Table 6.3 Net present cost (floor finishes)

Item	£ cost per m² (not discounted)	Year	Discounting factor	Net present cost per m² (£)
Option 1 Replace vinyl with vinyl sheet				
Initial cost	12.00	0	1	12.00
Annual cleaning	10.00	1	0.9524	9.52
	10.00	2	0.9070	9.07
	10.00	3	0.8638	8.64
	10.00	4	0.8227	8.23

Table 6.3 contd.

Item	£ cost per m² (not discounted)	Year	Discounting factor	Net present cost per m² (£)
	10.00	5	0.7835	7.84
	10.00	6	0.7462	7.46
	10.00	7	0.7107	7.11
	10.00	8	0.6768	6.77
	10.00	9	0.6446	6.45
	10.00	10	0.6139	6.14
Replacement	12.00	10	0.6139	7.37
Total net present cost per m²				96.60
Option 2 Replace vinyl with carpet tile				
Initial cost	20.00	0	1	20.00
Annual cleaning	2.50	1	0.9524	2.38
	2.50	2	0.9070	2.27
	2.50	3	0.8638	2.16
	2.50	4	0.8227	2.06
	2.50	5	0.7835	1.96
	2.50	6	0.7462	1.87
	2.50	7	0.7107	1.78
	2.50	8	0.6768	1.69
	2.50	9	0.6446	1.61
	2.50	10	0.6139	1.53
Replacement	12.00	10	0.6139	7.37
Total net present cost per m²				36.68

Component and material selection

When the design team have evaluated the broad construction options, presented these options to the client, and a decision has been reached, the design team will then need to make a detailed decision in relation to component selection. The choice of components used in a construction project will depend upon the function of the building and the 'building life' required by the building owner.

Life cycle costing techniques can be used to evaluate the whole life costs of particular components. As an example, in constructing an infants school a local authority has to decide upon the building material to be used for the cladding of external walls. The design team propose two alternatives, which are a metal clad external wall or a cladding comprising facing bricks. The building owner requires a 'building life' of 40 years and the discount rate used for the purposes of the life cycle cost plan is 5%.

The design team, in collaboration with the building owner, decide to restrict their choice of external cladding to the building to the following specifications:

(1) Corrugated or troughed sheet cladding
   Galvanised steel troughed sheets; BSC strip mill products; colourcoat 'Plastisol' finish

Wall cladding; vertical; fixing to steel rails with plastic headed self-tapping screws
0.7 mm scan rib 1000; 19 mm deep

(2) Brick facework
Facing bricks: PC £150.00/1000; in gauged mortar (1:1:6)
Skins of hollow walls; shallow recessed pointing one side as work proceeds
Half brick thick

Based on the adoption of a discount rate of 5% and the building maintenance manager's assessment that metal cladding would need to be replaced at 20 year intervals and that the brick facings would need a 50% repointing after 20 years and a full repointing after 40 years, the net present cost of these two options is as shown in Table 6.4.

Table 6.4 Net present cost (wall cladding)

Item	£ cost per m² (not discounted)	Year	Discounting factor	Net present cost per m² (£)
Option 1 Metal cladding				
Initial cost	25.00	1	0.9524	23.81
Replacement	27.00	20	0.3769	10.18
Replacement	27.00	40	0.1420	3.83
Total net present cost per m²				37.82
Option 2 Brick facing cladding				
Initial cost	33.00	1	0.9524	31.43
Repointing of 50% of facings	3.00	20	0.3769	1.13
Repointing of 100% of facings	6.00	40	0.1420	0.85
Total net present cost per m²				33.41

Property portfolio management

Life cycle costing can be used not only to evaluate options relating to new construction projects. The technique can also be employed to aid investment decisions for clients with existing property portfolios. The technique can therefore be used as detailed in the section on design choice earlier in this chapter, to compare options relating to an existing property or range of properties, with proposals for new investments in property to fulfil the client's objectives. This technique is usually termed 'investment appraisal'.

Investment appraisal

Investment appraisal may be defined as a systematic approach to investment decisions. The technique is based on the methods used in life cycle costing but investment appraisal is also concerned with a consideration of the non-financial benefits and disadvantages of each of the options considered.

An investment appraisal should normally adopt the following sequence:

- Define the objectives.
- Generate the options.
- Examine and compare the options.
- Evaluate the options.
- Consider the non-financial benefits and disadvantages of the options.
- Present the results and make a decision.

The first step in investment appraisal is for the client to define the objectives. The client's objectives may relate to the need to construct a single project or to a section of the client's property portfolio. The client's objectives will relate more to his needs in running an enterprise or providing a service, than whether new or existing buildings are to be used.

The second stage is the generation of a number of options that meet the client's objectives. In the case of investment appraisals relating to property portfolios, it is a useful technique to always assess a number of 'yardstick' options against which more considered options can be compared. For example, where a local authority is examining a portfolio of school buildings 'yardstick' options would be used to examine extreme courses of action. As an example, two 'yardstick' options could be adopted as follows:

Option 1 Status quo option
This option would be used where no new capital work whatsoever would be undertaken and the intention would be to maintain the status quo throughout the 'life of the investment' selected for the appraisal.

Option 2 The 'do everything' option
This option would envisage giving the design team a free hand to improve the property stock by a complete replacement of the stock with new buildings more fitted to the existing needs of the client in terms of space, design, and amenity standards.

It is suggested that these two extreme options would not normally present the best investment course for a building owner but they form the basis against which a number of more considered options can be measured.

The next step is to examine and compare the options. This examination will lead to an evaluation in financial terms of each of the options. This evaluation will be expressed in terms of a net present cost or value for each of the options. The net present cost or value will be the measure of the economic worth of an option against a defined 'life of the investment' and an agreed discount rate.

Although investment appraisal techniques are concerned with an evaluation of the economic worth of different options, clients will also be interested in the non-financial benefits or disadvantages of any particular option. The technique of cost benefit analysis can be adopted. However, it is recommended that, for the majority of cases, an investment appraisal will be made very much more meaningful to a client if the non-financial benefits and disadvantages associated with each of the options are given by way of a series of simple statements. An example of how this technique can be used is given in the case study later in this chapter, in the section on limitations.

Having completed an investment appraisal and assessed each of the options to be considered both in financial and non-financial terms, it is of paramount importance that the results are presented to the client in a clear and precise form. Most clients will not be interested in the mechanics of the calculations. They will want to see the bottom line figures giving the economic worth of each of the options clearly stated. They will also want to examine the considered non-financial advantages and disadvantages of each of the options. In most cases it will be appropriate in undertaking the investment appraisal to consider the non-financial benefits and disadvantages of each of the options to be evaluated with the client. These factors will influence his decision in choosing the preferred option.

MAJOR ISSUES

The role of the client

The client plays a crucial role in life cycle costing for new projects or projects relating to his existing property portfolio. Every project represents a significant investment by a client. The client in commissioning a construction or refurbishment project should become actively involved with the design team and ensure that his requirements are understood and satisfied in the options proposed by the design team. The chosen option may not necessarily be the most economic option but it will need to fit within the client's constraints in terms of capital and cash flow requirements.

The client will contribute to life cycle cost planning by providing relevant cost data. The client will also contribute by defining the life of the investment to be used in the life cycle cost plan, and the likely residual value of the asset when the client no longer has a use for it.

The relationship between the quantity surveyor and other professionals engaged by the client

In some instances a client may appoint a quantity surveyor to undertake life cycle costing on his behalf. However, major building clients will usually have other professionals such as bankers, accountants, valuation surveyors and architects, who he may expect to contribute to any life cycle costing or investment appraisal. The quantity surveyor will therefore need to have a flexible approach to life cycle costing and to provide the service required within the context of any given blend of professional skills that the client will bring to bear on the project.

The collection of data

Life cycle costing has been regarded by many quantity surveyors as being difficult to implement because of the difficulty of assembling reliable data. This issue is dealt with further in the next section.

Quantity surveyors undertaking life cycle cost planning should endeavour to establish the likely costs to be considered in any study in consultation with the client and by reference to the cost of any similar buildings operated by the client or by owners of similar projects. In the absence of good information systems maintained by clients the collection of data can be tedious and the quantity surveyor will need to draw on any published sources that are available. In the absence of published data the quantity surveyor will have to calculate costs from basic principles where appropriate.

LIMITATIONS

The client's awareness of the technique of life cycle costing

Many clients are unaware of the technique of life cycle costing and how it can help them to make better investment decisions. However, many other major clients and all public sector clients are aware of the techniques of life cycle costing and investment appraisal and are increasingly calling upon quantity surveyors to offer this service to existing property portfolios.

There is a clear need for the quantity surveyors to market their skills in the field of life cycle costing. The quantity surveyor combines the skills of cost planning and construction with the ability to assemble occupancy costs and apply discounting factors to the time-stream of costs and revenues to produce the economic worth of any particular option. The professional skills of the quantity surveyor therefore place him in a key position to undertake life cycle costing and investment appraisal work on construction projects. Quantity surveyors must therefore grasp this opportunity or it will surely be lost to other professions.

The reliability of data

Data for life cycle costing is uncertain because much of it relates to future costs that will be affected by inflation and other factors. However, much of the concern in relation to the reliability of data can be overcome by the use of the sensitivity testing of results where it is thought that a different ranking of options would be achieved by changing certain parameters.

The reliability of assumptions

The principal assumptions that need to be made in relation to any life cycle costing exercise are the life of the investment, the rate of discount to be used and the likely residual value at the end of the project. The discount rate will depend upon the issues referred to in an earlier section.

Once again, it is suggested that in cases where there are small differences between the ranking of some or all of the options considered then the major parameters of project life, discounting and residual values should be examined by sensitivity analysis to establish whether the ranking would be changed by an amendment to any of the major parameters.

CASE STUDY

Introduction

The case study relates to a life cycle costing of two options for the provision of a primary school by a local authority.

The local authority is currently educating 90 primary school children in a school built in 1890 capable of accommodating 210 children. Although the building is in sound condition the premises are costly to maintain in respect of energy costs, cleaning costs and building maintenance due to the large volume of the building and the excess floor area that is not required.

The local authority decided to undertake a life cycle costing to assess the feasibility of closing the existing school, disposing of the asset and building a new school to a floor area more in line with present demands on an adjacent site.

The local authority has decided to base the life cycle costing on a life of the investment of 20 years and to use a discount rate of 5%. The use of a discount rate of 5% implies inflation of future costs and values, and therefore all future costs and values included in the study are priced at today's prices.

Tables 6.5 and 6.6 respectively show the net present values of maintaining the existing primary school built in 1890 having a gross internal floor area of 1755 m^2 and building a replacement primary school having a gross internal floor area of 600 m^2.

The net present value of maintaining the existing primary school is £700 372 and the net present value of building a replacement primary school is £569 504.

It will be seen from the Tables that in the case of both projects some capital is required either in terms of a major refurbishment for the existing building or for constructing the new primary school. In the case of the replacement primary school the life cycle costing shows a credit for the sale of the existing site.

The operations costs for energy, rates and insurances, caretaking and cleaning have been shown as being constant throughout the life of the project. The replacement primary school has lower operations costs due to the smaller gross internal floor area and the smaller volumes associated with the proposed replacement school.

The intermittent maintenance which occurs at various cycles throughout the life of the project shows that the existing school will generate higher levels of intermittent maintenance.

The annual maintenance for both the existing school and the replacement

Table 6.5 Net present cost of maintaining existing school

MAINTAIN EXISTING PRIMARY SCHOOL (BUILT IN 1890)
HAVING A GROSS FLOOR AREA OF 1755 M2

	YEAR 0	YEAR 1	YEAR 2	YEAR 3	YEAR 4	YEAR 5	YEAR 6	YEAR 7	YEAR 8	YEAR 9	YEAR 10	YEAR 11	YEAR 12	YEAR 13	YEAR 14	YEAR 15	YEAR 16	YEAR 17	YEAR 18	YEAR 19	YEAR 20
CAPITAL COSTS																					
Major Refurbishment		50000	50000																		
OPERATIONS COST																					
Energy	7750	7750	7750	7750	7750	7750	7750	7750	7750	7750	7750	7750	7750	7750	7750	7750	7750	7750	7750	7750	7750
Rates/Insurance	10000	10000	10000	10000	10000	10000	10000	10000	10000	10000	10000	10000	10000	10000	10000	10000	10000	10000	10000	10000	10000
Caretaking/Cleaning	15900	15900	15900	15900	15900	15900	15900	15900	15900	15900	15900	15900	15900	15900	15900	15900	15900	15900	15900	15900	15900
MAINTENANCE INTERMITTENT																					
1) Building Elements																					
External walls						10000					10000					10000					
Roof						3000					3000					3000					
Internal Finishes						2000					2000					2000					
2) Service Elements																					
Sanitary Hot/Cold Water						2500					2500					2500					
Heating and Ventilation						3500					3500					3500					
Lighting Installation						1000					1000					1000					
3) Decoration																					
External								3000							3000						
Internal						3000					3000					3000					
MAINTENANCE ANNUAL																					
1) Building Elements	3500	3500	3500	3500	3500	3500	3500	3500	3500	3500	3500	3500	3500	3500	3500	3500	3500	3500	3500	3500	3500
2) Service Elements	5250	5250	5250	5250	5250	5250	5250	5250	5250	5250	5250	5250	5250	5250	5250	5250	5250	5250	5250	5250	5250
3) External Works	4000	4000	4000	4000	4000	4000	4000	4000	4000	4000	4000	4000	4000	4000	4000	4000	4000	4000	4000	4000	4000
RESIDUAL LAND VALUE																					-180000
TOTAL	46400	96400	96400	46400	46400	71400	46400	49400	46400	46400	71400	46400	46400	46400	49400	71400	46400	46400	46400	46400	-133600
DISCOUNT RATE	1	0.9524	0.907	0.8638	0.8227	0.7835	0.7462	0.7107	0.6768	0.6446	0.6139	0.5847	0.5568	0.5303	0.5051	0.481	0.4581	0.4363	0.4155	0.3957	0.3769
NET PRESENT VALUES	46400	91811	87435	40080	38173	55942	34624	35109	31404	29909	43832	27130	25835	24606	24952	34343	21256	20244	19279	18360	-50354
TOTAL NET PRESENT VALUE	700372																				

Table 6.6 Net present cost of replacing school

BUILD REPLACEMENT PRIMARY SCHOOL
HAVING A GROSS FLOOR AREA OF 600 M2

	YEAR 0	YEAR 1	YEAR 2	YEAR 3	YEAR 4	YEAR 5	YEAR 6	YEAR 7	YEAR 8	YEAR 9	YEAR 10	YEAR 11	YEAR 12	YEAR 13	YEAR 14	YEAR 15	YEAR 16	YEAR 17	YEAR 18	YEAR 19	YEAR 20
CAPITAL COSTS																					
Land	120000																				
Building		135000	135000																		
Credit For Sale of Existing site				-180000																	
OPERATIONS COST																					
Energy	7750	7750	7750	3000	3000	3000	3000	3000	3000	3000	3000	3000	3000	3000	3000	3000	3000	3000	3000	3000	3000
Rates/Insurance	10000	10000	10000	5000	5000	5000	5000	5000	5000	5000	5000	5000	5000	5000	5000	5000	5000	5000	5000	5000	5000
Caretaking/Cleaning	15900	15900	15900	11700	11700	11700	11700	11700	11700	11700	11700	11700	11700	11700	11700	11700	11700	11700	11700	11700	11700
MAINTENANCE INTERMITTENT																					
1)Building Elements																					
External walls															3800						
Roof															2000						
Internal Finishes										1000					1000						
2) Service Elements																					
Sanitary Hot/Cold Water											1000				1000						
Heating and Ventilation											3000				1500						
Lighting Installation											500				500						
3)Decoration																					
External						1100					1100					1100					
Internal								2000						2000							
MAINTENANCE ANNUAL																					
1) Building Elements	3500	3500	3500	750	750	750	750	750	750	750	750	750	750	750	750	750	750	750	750	750	750
2) Service Elements	5250	5250	5250	1500	1500	1500	1500	1500	1500	1500	1500	1500	1500	1500	1500	1500	1500	1500	1500	1500	1500
3) External Works	4000	4000	4000	2000	2000	2000	2000	2000	2000	2000	2000	2000	2000	2000	2000	2000	2000	2000	2000	2000	2000
RESIDUAL LAND VALUE																					-120000
TOTAL	166400	181400	181400	-150050	23950	25050	23950	25050	23950	23950	30550	23950	23950	25650	34850	23950	23950	23950	23950	23950	-96050
DISCOUNT RATE	1.00	0.9524	0.907	0.8638	0.8227	0.7835	0.7462	0.7107	0.6768	0.6446	0.6139	0.5847	0.5568	0.5303	0.5051	0.481	0.4581	0.4363	0.4155	0.3957	0.3769
NET PRESENT VALUES	166400	172765	164530	-134796	19704	19627	17871	18443	16209	15438	18755	14004	13335	12701	13107	16763	10971	10449	9951	9477	-36201
TOTAL NET PRESENT VALUE	569504																				

school are shown as being constant throughout the life of the existing school. In the case of the replacement school as the replacement will not be occupied until year 3 operations costs and the annual maintenance for the existing school will be required and these costs are used for years 0, 1 and 2 in the replacement school option.

The residual land values in both cases have been based entirely on site values it being assumed that the buildings themselves would not be readily saleable on the property market due to the specialist nature of the accommodation.

The life cycle costing of these two options shows that it would be more economic to build a replacement primary school disposing of the existing school, than to continue maintaining the existing primary school built in 1890. This is because it would require a sum invested now of £569 504 to build a replacement primary school and operate and maintain it over a 20 year life cycle. It would require a sum of £700 372 invested now to maintain the existing primary school over the same life cycle.

Clients will not usually make decisions based purely on the economic measure of an option evaluated by the use of life cycle costing techniques. They will need to take a number of other factors into account which will impinge upon the delivery of their service (in the case of a local authority) or the running of their business enterprise in the private sector. In the case of a local authority considering primary school education they will be taking a number of factors into account, which cannot directly be measured in economic terms.

Cost benefit analysis can be used to evaluate and weight the significance of the non-financial factors affecting different options in an investment appraisal study which uses life cycle costing techniques. However, in this case study the local authority asked the client team and the professional advisers to list the advantages and disadvantages of each of the options. These were given as a series of 'one-liners', which although brief, give the reader an indication of the major non-financial factors associated with both the options. Table 6.7 represents a typical summary of the report showing the net present value of the two options and the principal non-financial advantages and disadvantages associated with each of the options.

Table 6.7 Case study: Life cycle costing of two options for providing primary school education at somedistrict, anytown

Option number	1	2
Description of option	Maintain existing primary school (built in 1890) having a gross internal floor area of 1755 square metres.	Build replacement primary school having a gross internal floor area of 600 square metres.
Net present value of option (£000)	700	570
Advantages	• Avoids relocation of school	• Improved buildings and teaching environment • Aids curriculum development of science and computer education

Table 6.7. Contd.

Option number	1	2
Advantages (cont.)		• Improved outside play facilities • More welcoming environment for parents • Contributes to regeneration of area
Disadvantages	• Building difficult to supervise as on two floors • Poor buildings and teaching environment • Restricted opportunities of curriculum development of science and computer education • Inadequate outside play facilities	• Involves relocation of school

APPENDIX A COSTS AND VALUES

Below is a suggested rather than exhaustive checklist of costs and values categories to be considered. Items may need to be added and some may not be applicable to any particular project.

Capital costs

1.1 Land
1.2 Fees on acquisition
1.3 Design team professional fees
1.4 Demolition and site clearance
1.5 Construction price for building work
1.6 Cost of statutory consents
1.7 Development land tax
1.8 Capital gains tax
1.9 Value added tax
1.10 Furnishings
1.11 Other capital costs
1.12 Commissioning expenses
1.13 Decanting charges

Financing costs

2.1 Finance for land purchase and during construction
2.2 Finance during period of intended occupation
2.3 Loan charges (public sector)

Operation costs

3.1 Energy
3.2 Cleaning
3.3 Business rates
3.4 Insurances
3.5 Security and health
3.6 Staff (related to the building)
3.7 Management and administration of the building
3.8 Land charges
3.9 Energy conservation measures
3.10 Internal planting
3.11 Equipment associated with occupier's occupation

Annual maintenance costs, intermittent maintenance, replacement and alteration costs

4.1 Main structure
4.2 External decorations
4.3 Internal decorations
4.4 Finishes, fixtures and fittings
4.5 Plumbing and sanitary services
4.6 Heat source
4.7 Space heating and air treatment
4.8 Ventilating systems
4.9 Electrical installation
4.10 Gas installations
4.11 Life and conveyor installation
4.12 Communications installation
4.13 Special and protective installations
4.14 External works
4.15 Gardening

Occupancy costs

5.1 Client's occupancy costs

Residual values

6.1 Resale value — building, land and plant and equipment
6.2 Related costs — demolition and site clearance and disposal fees and charges

BIBLIOGRAPHY

There is now a developing literature on life cycle costing. The following references are selective, but will provide the reader with good coverage and a wide range of supplementary references.

Dell'Isola, P. E. and Kirk, S. J. (1981) *Life Cycle Costing for Design Professionals*. New York: McGraw Hill.

Flanagan, R. and Norman, G. (1983) *Life Cycle Costing for Construction*. London: Surveyors Publication, RICS.

Flanagan, R., Norman, G., Meadows, J. and Robinson, G. (1989) *Life Cycle Costing: Theory and Practice*. Oxford: BSP Professional Books.

Norman, G. (ed.) (1987) *Management and Economics: Special Issue*, Vol. 5. London: E. & F.N. Spon Ltd.

Royal Institution of Chartered Surveyors (1986) *A Guide to Life Cycle Costing for Construction*. London: Surveyors Publications, RICS.

Royal Institution of Chartered Surveyors (1987) *Life Cycle Costing: A Worked Example*. London: Surveyors Publications, RICS.

Chapter 7

Expert Systems Methodology

GEOFFREY H. BROWN, *Monk Dunstone Associates, London*
and JOYCE E. STOCKLEY, *Surveying Division,*
University of Salford

INTRODUCTION

Expert systems, an application of artificial intelligence, are growing in popularity inside business and industry, being seen as problem solving tools for very practical ends.

The development process for expert systems is different from that for standard software. The majority of the development time required for standard software is taken up in coding. The reverse is true of expert system development where the majority of the time is spent in planning and, in particular, in deciding what knowledge should be encoded into the system.

The aim of this chapter is to provide guidelines to practitioners in the quantity surveying profession on how to build expert systems, and since the acquisition of knowledge in such systems is often quoted as a major difficulty it is this aspect of system development which will be considered in detail.

An expert system definition

A working definition of an expert system has been given by Goodall (1985) as:

'a computer system which uses a representation of human expertise ... to perform functions similar to those normally performed by a human expert ... and operates by applying an inference mechanism to a body of specialist expertise represented in the form of knowledge.'

The broad stages in the development of an expert system

The standard methodology for building an expert system involves the following issues:

- setting the system's objectives;
- knowledge acquisition;

- knowledge representation;
- system testing/validation.

These may be addressed by considering development in terms of six steps:

(1) start
(2) skeleton system
(3) demonstration system
(4) working system
(5) usable system
(6) commercial system

These steps have been set out in this way to give the development process some structure for deciding what facilities to attend to, and when. It should be noted that, in practice, these steps overlap.

Step one: start
Step one involves the consideration of uses, users, roles and benefits of the expert system, and the definition of objectives, which all assist with the identification of knowledge requirements.

Basden (1983) has classified expert systems into the following seven roles:

(1) *Consultancy*. For use by a non-specialist to obtain specialist advice or other forms of help in accomplishing some task. Usually the user will be semi-skilled or sometimes a novice.
(2) *Checklist*. A reduced version of the consultancy role, where the emphasis is not on the result generated but rather on ensuring that all relevant questions have been asked and their answers recorded.
(3) *Training*. May be aimed at either initial training or for improving the expertise of people with a general rather than deep understanding of a domain. This role contains detailed explanation and 'what if' capabilities.
(4) *Knowledge refinement*. May be used as a repository of knowledge in the given domain which can be used by experts to highlight weaknesses in current understanding.
(5) *Communication*. May be used to make 'private knowledge' more accessible and capture specialist knowledge which might otherwise be lost through job changing or retirement.
(6) *Programmed system*. To express knowledge in a more understandable and easier-to-read form than conventional computer systems.
(7) *Demonstration*. To convey the concepts, capabilities and constraints of expert systems to an unaware public.

By defining the systems role in this way a broad indication is given of the class of knowledge needed. For instance, problem solving knowledge would be needed for a system with a consultancy role whereas knowledge on explaining concepts would be more important to fulfil a training role.

Each role is likely to yield certain benefits so it is important to establish which are expected early on. This will guide the system developer on the knowledge to encapsulate since it affects the level of accuracy provided, the content of reports, etc.

Typical benefits are as follows:

- *Consultancy role.* Offers greater speed in coming to a decision, easier access to expertise and greater reliability and consistency.
- *Checklist role.* Aids consistency, reliability and record keeping.
- *Training role.* Imparts good practice to the trainee.
- *Knowledge refinement.* Sharpens up the available human expertise.
- *Communication role.* Improved ability to communicate complex and judgmental expertise.

In considering the roles and benefits which the system should have, the likely effect of the system should also be taken into account at an early stage. This may be done at four levels:

- The level of the user: will the system be seen as a threat or nuisance and therefore not used?
- The level of the organisation: will organisational changes be required to run the system efficiently?
- The wider social level: will there be legal consequences, or detrimental effects on others outside the organisation?
- The environmental level: will it adversely affect the global environment?

These considerations help to highlight potential areas of trouble so that they can be allowed for in designing the way the system is built. Once the proposed role for the expert system has been identified, items or goals whose value or probability are to be evaluated may be defined.

Step one also involves the selection of the most appropriate tool to build the expert system. The most basic tool is the programming language. If a high level declarative language such as Prolog or Lisp is used it will involve building both the knowledge representation mechanism and the reasoning mechanism from scratch, which gives great flexibility in tailoring them exactly to the applications under consideration.

The usual alternative is to use an expert system shell which contains both a ready-made inference mechanism and some form of knowledge representation scheme. The benefit of using a shell is the saving in time and effort, and thereby cost, taken to build the system compared with using a high level language which may involve developing facilities that are available in existing shell systems.

The planning involved in this first stage often determines the overall success of the project.

Step two: skeleton system
A skeleton system may be constructed after the roles and objectives have been

set. It can incorporate sufficient knowledge to act in approximately the right way but need not give accurate answers. Its purpose is to give the developer some idea of the domain and what can be expected from the system.

Step three: demonstration system
This is the skeleton system 'fleshed out' into a demonstration system to provide more accurate results, but mainly to establish the feasibility of continuing with the development. This demonstration system may be achieved some 3 to 6 months into the project, by which time the developer has a reasonable idea of the complexity of the knowledge and the users, and the capability of the technology. It is at this stage that the majority of the knowledge acquisition is undertaken, and at which much of the knowledge is implemented in software (or coded into the appropriate representation mechanism).

The ease with which knowledge may be coded in software, and the time it takes to do so, depends on a number of different factors such as the tool selected to build the system, the familiarity of the system builder with the development tool and the complexity of the knowledge to be represented. The emphasis of this chapter is on knowledge acquisition rather than representation, but it should be noted that the task of coding can be very time consuming.

Step four: working system
The working system is a development of the demonstration system, but validated and debugged so as to generate accurate results. The following procedure is suggested at this stage:

● The elimination of 'bugs' such as rules applying in the wrong circumstances. These can usually be quickly spotted by experts seeing discrepancies between their own reasoning and that of the system.
● The identification of gaps. Since a human's knowledge is never complete it would be sufficient for the knowledge base to contain the correct spread of knowledge to perform reliably and robustly in even unpredictable circumstances.
● The evaluation of the system's performance with the proposed user population. This involves testing the system on a set of real tasks to check that the appropriate level and quality of decision support has been provided to enable the system to fulfil its proposed role effectively.

In theory the system could be used in earnest at this stage, although questions and reports are still likely to be poorly worded and enhancing facilities such as links to a database may not yet be incorporated.

Step five: usable system
The usable system is the working system with additions to improve its 'use' such as a help system and the facility for altering screen colours. A usable system can be used by those who are sympathetic to it to obtain real business benefit.

Step six: commercial system
A full commercial system results once the wording of reports and questions has

been perfected, and comprehensive support is available to the user through clear user manuals, etc.

KNOWLEDGE ACQUISITION

It can be seen from the definition of an expert system provided above that the effectiveness of its operation is dependent on the quality and scope of the knowledge contained within it, and how adequately that knowledge is represented. This is why the process of acquiring knowledge and representing it needs careful consideration. Most of the knowledge incorporated into a system is acquired during the first three steps of its development as noted above.

The objective of knowledge acquisition

Knowledge acquisition is the transfer and transformation of problem solving expertise from some knowledge source to a program. The potential sources of knowledge include human experts, textbooks, databases and the system developer's own experience.

The acquisition of 'public' knowledge, i.e. published facts and theories typically found in textbooks and the like, is relatively straightforward. The acquisition of 'private' knowledge which is usually stored in the human mind involves using knowledge elicitation techniques, i.e. the extraction of knowledge from the human expert.

Advantages of knowledge elicitation

As suggested by Kidd and Welbank (1984), knowledge elicitation can be a very valuable process, as apart from the direct benefits obtainable from a working expert system the following indirect benefits may also result:

- The actual process of knowledge elicitation can help sharpen an expert's thinking, by forcing him to reconsider the rules he uses and the order in which he works.
- The process of extracting and codifying scattered or disorganised knowledge can reveal that certain key facts are missing, or uncover new previously unrealised facts.
- Since human expertise is both fragile and transient the ability to archive the expert's knowledge improves its accessibility.

Common knowledge elicitation techniques

The term 'knowledge engineer' is commonly used for those individuals who specialise in developing expert systems. Two standard methods are relied on for extracting knowledge from an expert and building it into a system:

(1) the knowledge engineer engages in intense interview with the expert;
(2) the knowledge engineer becomes an expert himself, relying on introspection to articulate the requisite knowledge.

Expertise is primarily a skill of recognition, of 'seeing' old patterns in a new problem. Olson and Rueter (1987) suggest that experts organise the information in their minds in a highly structured manner, using a variety of kinds of knowledge structures dependent on the type of information.

The variety of kinds of knowledge structures suggested by them includes the following:

- Simple lists, e.g. the days of the week, months of the year.
- A table, e.g. calendar appointments.
- A flow diagram, e.g. a decision tree for representing the routing of telephone messages to the people who can handle them.
- In hierarchies, e.g. relationships.
- In networks, e.g. richly connected language associations.
- As physical space, e.g. room arrangements or maps.
- As a physical model, e.g. information about a device's internal components and how they are causally related.

Knowledge may therefore be held in many different representations, procedural, causal, classificatory, conceptual, etc., each suitable for a particular kind of reasoning or retrieval.

It is essential that the knowledge engineer appreciates that human expertise is comprised of such varied types of knowledge. If, in extracting expertise, only the problem solving knowledge which is often expressed in terms of rules of thumb is focused on, then there is a danger that expert systems will only be considered in terms of facts and rules, and the importance of the other types of knowledge which would make them more effective would be overlooked.

Methods for revealing what experts know may be classed as either 'direct' or 'indirect'.

Indirect knowledge elicitation methods are more applicable to those experts who are unable to explain their reasoning, so they collect other behaviours such as recall or scaling responses from which inferences can be drawn about what the expert 'must have known' in order to respond in that way.

Indirect methods include: multi-dimensional scaling; Johnson hierarchical clustering; general weighted networks; ordered trees from recall; repertory grid analysis. Practitioners who wish to see a detailed appraisal of indirect methods of knowledge elicitation should refer to the paper by Olson and Rueter (1987).

Experts in the field of quantity surveying are expected to be able to express their knowledge directly, to substantiate their advice to the client. It is therefore appropriate to concentrate on the use of direct knowledge elicitation methods with these experts since they ask the expert to report on knowledge which can be directly articulated.

Direct methods of knowledge elicitation include: interviews; questionnaires;

simple observation; thinking-out-loud protocols; interruption analysis; drawing closed curves; inferential flow analysis.

Interviews

In conversation the expert reveals the objects considered, and the processes gone through, in making a judgement, solving a problem or designing a solution. Interviews are the most common method for eliciting knowledge from an expert. They are free form, where experts can generate information in the order and detail they wish. The following guidelines may be followed to make interviewing more efficient:

(1) enlist the expert's co-operation;
(2) ask free-form questions at the start, narrowing in specificity as the interview process progresses;
(3) do not impose your own understanding on the expert;
(4) limit the sessions to coherent tasks, recognising fatigue and attentional limits.

These aspects are discussed in more detail later, in the section on knowledge elicitation.

Interviews are very time-consuming and there are limits to the completeness of information which can be obtained in this free form style. This problem may be eased by introducing questions of other styles, such as focusing on a particular case in detail, called the 'critical incident technique'. This concentration on a case helps to elicit particular descriptions, rules and objects, which can be examined for their generality in later sessions. By asking for symptoms and characteristics one elicits features, while asking for evidence elicits inferences.

Questionnaires

These consist of cards or pieces of paper printed with standard but open-ended questions, rather than the kind used in survey research. Questionnaires are an efficient way of gathering information, where the expert can fill them out in a leisurely and relaxed atmosphere. They can be particularly useful in discovering the objects of the domain, in uncovering relationships and perhaps in determining uncertainties.

Observation of the task performance

The expert is watched by the knowledge engineer as he tackles a real problem. The knowledge engineer must decide how to record the expert's performance, e.g. to watch, take notes, try to follow the thinking process on the fly; or to videotape the process for later review with the expert. The former method suffers from time pressure and observer bias, and the latter relies on the expert's less than perfect ability to recall the reasons underlying his performance.

Protocol analysis

In performing a task the expert is asked to think out loud, explaining as the task progresses 'what the goal is', 'what the method to be used is' and 'what is the expert considering'. The performance is, as with the method of simple observation, recorded by the knowledge engineer.

The goal in obtaining a protocol lies in identifying the kinds of objects the expert sees/considers, the relationship that exists between the protocols and the kinds of inferences drawn from the relationships seen.

The advantage of this method is that there is no delay between the act of thinking something and reporting it. It takes up a minimum of the expert's time, and is most suitable for those tasks where an expert would naturally make inferences to himself or identify salient features of the objects in the situation as he proceeded. However, it may cause the expert to adopt a more systematic approach than normal, and is unlikely to be suitable for tasks where no natural verbalisation would normally occur such as in composing music.

Interruption analysis

This method also involves observing the expert's performance, but allows him to proceed with a task without thinking aloud, until the point when the knowledge engineer no longer understands the expert's thought processes. At this point the expert is interrupted and asked to explain in detail why he is working in that manner. This method trys to capture the focus of attention and the inferences drawn for the features noticed at the point of interruption, but it may break the expert's train of thought.

Drawing closed curves

The expert is asked to indicate which of a collection of physical objects 'go together', to draw the related objects in a closed curve. This technique is applicable to any spatial representation such as a position on a game board. (It was considered the least appropriate of all the direct methods in application to quantity surveying experts.)

Inferential flow analysis

This is a variant on the interview where answers to particular questions about causal relationships are used to build up a causal network among concepts or objects in the domain of expertise. Key objects in the domain are listed and the expert is asked about the relationship between two of them.

Method summary

Interviews, questionnaires, observation of the task performance, protocol analysis and interruption analysis attempt to reveal the contents of the thought processes during the solution of existing problems. They highlight the vocabulary the expert uses to identify the objects and their relationships, and the kinds of inferences drawn. These methods are free of assumptions about the form of the relationships among the items, be they lists or tables or networks or physical space, thus giving the knowledge engineer a chance of finding any kind of information.

The last two methods of direct knowledge elicitation are specifically designed to elicit one particular kind of information over another, where drawing closed curves explicitly illuminates the relationships among objects in the problem space and inferential flow analysis displays the inference chains experts may use to reach conclusions.

The knowledge engineer must make judgements of the suitability of a method for knowledge elicitation to the kinds of knowledge the expert is assumed to possess. It is likely that the best results will be obtained using as many of the different techniques available as possible in a carefully tuned combination.

EXPERT SYSTEM METHODOLOGY: AN APPLICATION

Details of the RICS/ALVEY Research Project

In 1983 the Alvey Programme was started. It was a 5 year initiative, backed by government and industry, which aimed to mobilise the UK's technical strengths in information technology (IT) and thereby improve the country's competitive position in the world's IT markets.

This initiative included promoting awareness of intelligent knowledge based systems (IKBS) which incorporate expert systems and are now widely acknowledged to be of crucial importance to the future application of computers through their ability to handle the sort of high level functions that require a degree of intuitive judgement.

Part of the awareness programme involved the setting up of community clubs in various industries, to commission the development of expert systems in relevant applications in order to demonstrate their potential use in those particular industries.

The Quantity Surveyors Division of the Royal Institution of Chartered Surveyors (RICS) was invited by the Alvey Directorate to set up the RICS Community Club, which it did in collaboration with the University of Salford. The aim of its research project was to undertake the development of an expert system relevant to the specific area, or domain, of quantity surveying.

A period of eighteen months was allotted to the development of the system, and two knowledge engineers and a chartered quantity surveyor were employed to build it, under the direction of Professor Peter Brandon at the University of Salford. The work undertaken was referred to as the RICS/ALVEY Research Project. A full account of the RICS/ALVEY Research Project is provided in the published report by Brandon et al. (1988).

Reasons for employing knowledge elicitation techniques within this project

Due to the nature of the research project and the size of the potential user group, i.e. the entire practising Quantity Surveyors Division of the Royal Institution of Chartered Surveyors, the RICS formed a representative user group. This comprised members of twelve quantity surveying organisations drawn from private practice, contracting, central and local government. The group's main responsibility was to provide the knowledge on which to build the system, thus involving the application of knowledge elicitation techniques to obtain it.

THE RICS/ALVEY RESEARCH PROJECT: MAJOR ISSUES

Setting the systems objectives

A series of discussions were held with the user group to identify the likely users and the role of the system, and thereby the system's performance requirements. The 'top down' design process suggested by Attarwala and Basden (1985) was applied to assist with determining this sort of criteria, i.e.:

- Who will the likely users be?
- What role will the system fulfil?
- What benefits can be expected, and for whom?
- What effect will the system have?

The users, in broad terms, were chosen to be chartered quantity surveyors since its development was commissioned by the RICS Community Club.

Views on the likely users within the QS profession varied between a system for use by:

(a) The expert, in which case it would take on the role of an intelligent prompt checklist.
(b) Those with less experience in the profession, in which case it would take on a training role.

The decision about role was influenced by the benefits expected from the system and the time constraints on the period available to build it. Designing the system for use by an expert alone would not demonstrate the supportive role which an expert system can perform. However, designing the system to fulfil a training role would require high levels of explanation and be very time consuming. It was therefore decided to aim the system for use by 'middle management' surveyors, where a certain level of familiarity with terminology and knowledge of standard practice could be assumed. The role for the system was therefore, one of decision support where the benefits envisaged were to improve the consistency of advice given to the client, and enhance the quantity surveyor's role as lead consultant.

The development of the system was commissioned to demonstrate the potential for using the technology within the profession. Consideration of system requirements therefore involved satisfying criteria related to the technology as well as the 'goal' objectives defined. The subject area chosen needed to illustrate the main feature of the expert system technology, i.e. to be one where judgemental functions would normally be undertaken by a human expert.

System credibility depended on the subject area addressing a real world problem and for practical reasons it had to be one where knowledge would be relatively scarce, and so demonstrate a support role, but attainable enough to allow a useful system to be produced.

The subject area chosen to satisfy all of these criteria was the strategic planning of construction projects. It was envisaged that the system would provide the framework to translate information on client needs and wants into a strategy concerned with the four main issues to be evaluated at the concept stage of a construction project. The four issues to be considered within the subject area were initial budget, procurement selection, time duration forecasting and development appraisal, and these formed the 'goals' for the system.

Knowledge acquisition

Initially, as much as possible was learnt about the domain and the task by reference to textbooks, documents, papers and general discussion with members of the quantity surveying profession.

Mindful of the danger of focusing only on the elicitation of problem solving knowledge, which is often expressed in terms of rules of thumb, the elicitation techniques adopted included consideration of the idea presented by Basden and Attarwala (1986) that human expertise can be presented as a model comprising understanding and experience in a circular relationship.

This model, represented in Figure 7.1, illustrates (as a portrait rather than a photograph) how after a period of experience the expert is able to construct rules of thumb by combining four types of knowledge, i.e.

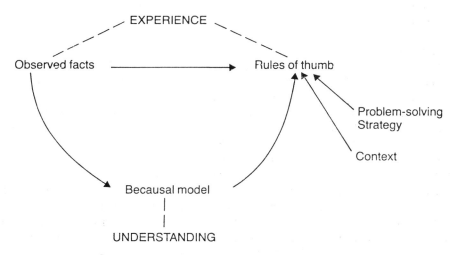

Figure 7.1 Model of expertise

(1) Understanding: of the domain.
(2) Problem solving strategy: based on intuition, past experience, preference and different problem solving skills.
(3) Context: a careful consideration of context taking account of only relevant factors and ignoring irrelevant ones.
(4) Observed facts: of a situation.

A mixture of informal interviewing and protocol analysis was used with the aim to elicit understanding, context and problem solving strategy knowledge relevant to the 'goals' to be evaluated.

Some structure was imposed on the process of building the expert system by asking the expert the following questions:

(1) Why?
(2) What else?
(3) When not?

'Why?' helps to separate out the different types of knowledge and is used primarily to identify understanding, although it can generate extra items. Understanding is often expressed as a causal model or an implication network of relevant items. Experts seem to find it more natural to give explanations of their advice in terms of cause and effect than attempting to articulate the knowledge they have.

'What else?' helps to focus the expert's mind on one particular item and any items immediately affecting it or affected by it.

'When not?' helps to generate context knowledge in identifying conditions when a relationship may be inoperative.

By focusing the expert's mind and helping to unearth 'forgotten' pieces of information this approach contributes to the speed and ease of building a system, and ensures a more complete knowledge base. The separation of the types of knowledge should also enable understanding to be used to reduce the likelihood of disagreements amongst experts and give more acceptable explanations to the user, since it is objective and not dependent on the individual's circumstances or approach to problems.

Knowledge representation

It is usually the system architecture or shell selected to build it which dictates the knowledge representation system to be used. By considering how the task is performed, what basic strategies are used and what may influence their use, etc. the system developer can attempt to formulate a model of the problem-solving process and match it to the most appropriate tool.

It was decided that the most appropriate tool to build the system for the research project was a commercially available shell, since the application was to establish awareness of the technology rather than deal with a specific problem which might require a more 'tailored approach'. In this way it was hoped to take full advantage of the savings in time, effort and cost associated with their use.

The shell selected to build the system was one which could cope with uncertainty. The shell was called Savoir, and it allowed the program to be written in terms of probability and thereby reflect the sort of imprecise and unknown project particulars typical at the concept stage of a project.

As well as incorporating the knowledge elicited into the Savoir shell, 'box and arrow' or inference net diagrams were adopted to express it as a network comprised of items and relationships, where an item represents a fact or numerical value and a relationship between two items means that some change in one item causes or implies a change in the other. In this way the network formed showed the causal mechanisms that contributed to the goal concerned.

To construct the network it was necessary to:

(1) Establish the major causal factors relevant to the goal (or the item or fact to be evaluated by the system). This may be achieved by asking the expert the questions 'why?', 'what else? and 'when not?'. When relationships were not detailed enough and more information was required, they may be expanded by finding out:

  (a) What is evidence for the item?
  (b) What is the item evidence for?

(2) Group the factors together and draw arrows among them and from them to the goal. An important aspect of this process is for the knowledge engineer and expert to establish the correct weighting of the factors. For instance, a relationship may exist between an item (A) and two other items (B and C). The relationship may be expressed as a rule which says that A depends on B and C, but B is more important than C. In representing this rule it is important to describe the strength of the relationships, and this may be achieved by using 'weights of evidence' to show the value of different items. The Savoir shell uses Baye's rule to derive the value of factors and defines the weights of evidence in terms of the odds of the item being true if the factor exists or not. The information on weighting may be obtained from the expert by asking the following questions:

  (a) In how many cases out of 100 would the item be true if this factor exists?
  (b) In how many cases out of 100 would the item be true if this factor does not exist?

(3) Add the questions that would be put to the user to establish these intermediate items, thus forming a chain of reasoning.

A detailed example of an inference net diagram is given in Figure 7.5.

System testing/validation

Three approaches were considered for the validation of the system. These were informal validation, structured experiments and comparative validation.

Informal validation involves a number of individual experts testing the system, at different periods during development, to establish the following:

- their agreement with the results provided by the system;
- their reactions to the questioning and screen presentations;
- their opinions on the general usability of the system.

Structured experiments are where test cases are presented for both experts and the system to provide advice, so that the approach and results of each may be compared.

Comparative validation involves testing the system on real world projects for which data is known and can be compared with that produced by the system.

The main approaches used were informal and comparative validation. Feedback from the testing undertaken allowed knowledge refinement to take place in an iterative fashion.

LIMITATIONS OF AN EXPERT SYSTEM METHODOLOGY

The suitability of the proposed domain should be considered before starting development, to ensure that the knowledge is readily available, and not subject to fast and frequent changes.

If a commercial shell is selected as the best tool to build the system care must be taken not to concentrate only on collecting knowledge that 'fits' within that shell. As recognised by Berry and Broadbent (1986) such 'shoe horning' of knowledge into shells results in the production of an ineffective system.

Because of the wide choice of expert system shells currently available it is recommended that before making a final selection, where possible, advice should be sought from system developers who already have experience in using the shell under consideration. The opportunity to capitalise on the experience of others helps to reduce the risk of a new development foundering through an incorrect choice of shell.

Expert systems are still limited in many ways, with certain kinds of knowledge remaining untouched by existing knowledge acquisition techniques. The extent to which such expertise may be captured in the future is still a matter for research. It is recognised there are many factors which influence the advice given by a human expert which are currently beyond the scope of expert system technology. For instance, it has difficulty handling visual information, such as the nature of a site and building shape and form. The human expert copes with these matters by a visual assessment of the site and codified information such as maps and drawings. To handle this within the expert system would involve a huge amount of additional knowledge, far beyond the return which its incorporation could provide.

Determining the boundaries of the knowledge to be incorporated within a system can also present problems. For instance, knowledge of client preferences and prejudices cannot realistically be incorporated into a system designed for use with numerous clients. This is the sort of specialised or local knowledge which individual users should be able to input 'manually'.

The effectiveness of the many knowledge acquisition techniques which exist depends on the situation concerned and the choice of knowledge representation method. Kidd and Welbank (1984) suggest that the problems encountered in the process of eliciting information may be due to the following:

- Human knowledge is very complex, and can be messy and ill-formulated.
- Humans find it very difficult to articulate what knowledge they have and how they use it to solve problems.
- The more expert someone becomes at a task the more 'unconscious' his knowledge becomes.
- The data resulting from knowledge elicitation is in the form of the expert's verbal comments or actions. These need careful interpretation to extract implied knowledge.
- The existing knowledge elicitation techniques are often of limited applicability.

Recognition of these problems is the first step to overcoming them. It is important that each expert system be considered for its particular requirements and a well-tuned mixture of techniques employed to obtain the richest collection of information possible about the knowledge used by experts. The ultimate success of a system depends on whether it contains sufficient knowledge to carry out the task and how accurate that knowledge is.

INTERPRETATION OF RESULTS

It is important that the performance of the system is rigorously tested against that of the human expert in order to verify the robustness, consistency and appropriateness of the knowledge base contents and the adequacy of the tool chosen to build it.

A good user interface to the system will help to encourage feedback from the experts participating in its testing and will improve its acceptance by users once development is complete.

Comprehensive reporting and explanation facilities make the expert system transparent to the user, enabling them to check on the assumptions made and the reasoning adopted.

By including facilities within the system which enable the user to interact with the final advice generated, e.g. overriding the assumptions made, etc. it allows a reliable but probably imperfect model to be improved by the knowledge of the user.

Because expert systems are designed to simulate the thinking process of the expert, following their logic, the user can identify with the process.

All of these factors speed up user confidence in the system, and its technology, making it more acceptable than many conventional programs. The consistency of approach in expert systems is likely to be greater than that of the expert. The system can default to the expert knowledge incorporated within it to solve the problem, and as it does not suffer the problems of human fallibility it is unlikely to make the same order of mistakes.

CASE STUDY

The aim of this case study is to provide a brief description of the approach taken in building the initial budget module program, which was developed as part of the RICS/ALVEY Research Project, together with an extract of an elicitation session.

As mentioned earlier (in this chapter), the overall subject choice for the expert system developed under the RICS/ALVEY Research Project was the strategic planning of construction projects. Four main issues, or modules, were considered which were interlinked through a common projects database. Conceptually the system looks like the diagram shown in Figure 7.2. The original 'goal' or objective for the initial budget module was defined as the cost of the building.

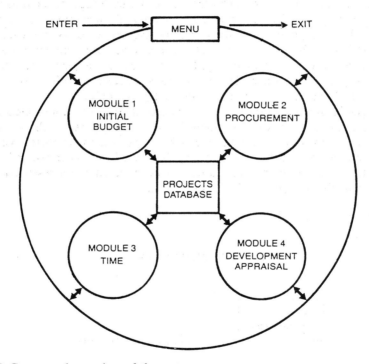

Figure 7.2 Conceptual overview of the system

Development approach

In deciding on the approach to take for the development of this module consideration was given to the methods usually adopted by quantity surveyors in setting an early budget.

Three methods were identified, the menu, the matching and the aggregate modelling approaches. The menu approach relied on the existence of a menu of guide costs for various building elements, from which a suitable cost could be selected. This provided the means for a quick estimate but would not be suitable for unusual buildings and was therefore considered to be of limited use.

The matching approach consisted of selecting, either from human memory or a database, a previous building project similar to the one in hand whose cost was known. This known cost could then be varied to take account of differences in shape, size, location, fittings, etc. The expertise required for the matching approach involved the identification of differences between buildings which had a significant cost implication. Since computers would have had difficulty in gauging similarities and differences between historic record and the project under consideration this approach was not considered suitable as a model to build the budget module on.

The aggregate modelling approach used a statement of client needs to 'design' and cost out, using unit rates, each element of the building, taking into account a suitable level of quality. This approach depended on the availability of a database of prices and knowledge of how to select among the options for each type of element. The modelling approach would normally be very time consuming for the quantity surveyor to employ manually. However, current expert system techniques could easily perform the complex calculations associated with it, so it was this approach which was used in building the budget module.

The stages in the aggregate modelling approach may be represented as Figure 7.3. The process is more detailed than Figure 7.3 suggests, as there are several types of design information, e.g. size and shape, quality factors, construction form, complexity and fitting out. Costing is carried out by means of elemental unit quantities and element unit rates. A more detailed representation of the aggregate modelling approach is shown in Figure 7.4.

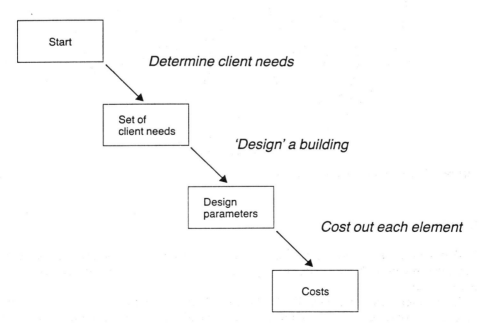

Figure 7.3 The stages in the aggregate modelling approach

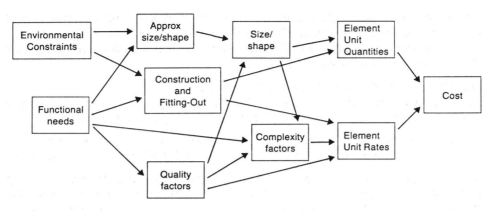

Figure 7.4 The aggregate modelling approach in more detail

A useful means of developing the system from this stage was found by considering element types in more detail to determine the factors affecting their selection and quality.

Knowledge elicitation

Before undertaking the knowledge elicitation sessions a published set of options for each element was obtained. This was then used as a basis for the subsequent interviews held with a number of experts.

The example which now follows deals with consideration of the roof element. Table 7.1 is a sample of the published data produced in the *Architects Journal*, 14 May 1986.

Table 7.1 Sample of published data

Type of roof	Rate (£/m²)
Pitched roof, tiles, with insulation	45 to 62
Pitched roof, composition slates, insulation	56 to 79
Pitched roof, Welsh slates, insulation	68 to 90
Mansard roof, slate-faced, insulation	79 to 135
Flat roof, timber, insulation	45 to 62
Flat roof, slab, insulation	62 to 73

The expert was shown this published data and then questioned about the element, and the following is an example of how the interview proceeded:

Question: 'When would each type of roof be used?'
Answer: 'Roof type depends mainly on the number of storeys, whilst the covering depends on quality.'

This information gave links between the number of storeys and the various roof types, and from roof type and quality to roof rate. From experience and observation it is recognised that some tall buildings have pitched roofs, whilst this is unlikely for a two-storey curtain walled building.

Question: 'What else determines the roof type?'
Answer: 'Architectural style, fashions for flat and/or mansard roofs which vary across the country and the need to let light into neighbouring buildings.'

Question: 'What else determines the rate, besides roof type and quality?'
Answer: 'The complexity of the roof and the need for wide span affect the rate.'

Interviews progressed in this manner and the whole knowledge base was built up in this way. As mentioned earlier, in the section on knowledge representation, the knowledge was both coded into the Savoir shell, to produce the expert system, and represented on inference net diagrams. These inference net diagrams proved to be very useful in subsequent elicitation sessions as they provided a rapid means of checking the information recorded to date. An example of the inference net diagram built up for roofs is shown in Figure 7.5.

ANALYSIS OF THE CASE STUDY, RELATED TO KNOWLEDGE ELICITATION

Interpretation of information

It is important to carefully consider verbal data elicited from an expert, to ensure that it is given the correct interpretation. The case study example in the previous section shows that some of the knowledge elicited in the development of the budget module concerned architectural fashion. To avoid distortion of the advice generated, it was therefore important to carefully consider whether such knowledge should be encapsulated, or not, as the system was only due to be used commercially some three years after the system's development.

To overcome this particular problem the fashion element was made explicit, so that if it changed it would be a simple matter to alter it. Some understanding of fashion was also sought. For instance, the current dislike of flat roofs is partly a reaction to their over-use in the 1960s and their tendency to leak. Since the dislike of flat roofs will probably be valid for some time to come, until new flat roof technology has been tried and tested to remove the tendency to leak, it was considered that this knowledge should be incorporated.

The use of an inference net diagram provided a good medium of communication between the knowledge engineer and the expert, to facilitate a rapid check on correct interpretation.

Knowledge elicitation notes

The following practical factors related to knowledge elicitation were identified during the course of the RICS/ALVEY Research Project:

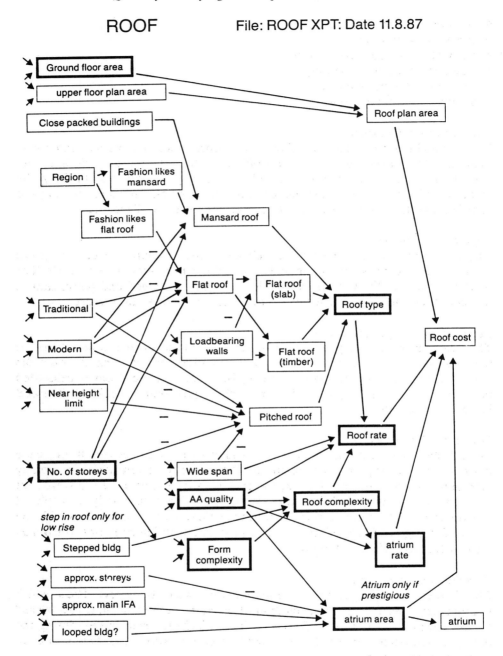

ROOF   File: ROOF XPT: Date 11.8.87

Figure 7.5 Inference net diagram for roof

- assessment of the expert;
- expert co-operation;
- recognition of topic exhaustion;
- selection of interview setting;
- appointment time selection;
- multiple expert situations.

Assessment of the expert

It is essential that the expert is enthusiastic. Before proceeding with an interview, therefore, the knowledge engineer should assess how the expert is feeling, and engage his interest and support gradually. This may be done by clarifying the objectives of the interview, summarising progress since any previous meeting and focusing attention using a visual aid such as inference net diagrams.

Expert co-operation

Any reluctance by the expert to participate in the development of the system should be overcome as early as possible if a useful contribution is to be made. The intended usage of the knowledge should be clearly explained and permission to document the interviews by using a tape recorder obtained in advance.

Recognition of topic exhaustion

The preparation of likely interview topics and questions in advance helps to maximise the use of an expert's time, and minimises their moving off 'onto the wrong track'. It was important to recognise when a topic became exhausted in order to be able to judge whether further questioning would elicit useful information, or not.

Selection of interview setting

The selection of the setting for an interview may influence its success. If the venue is away from the expert's own environment there may be difficulties in actually attending, due to pressure of work, the travelling time may present a constraint on interview duration, and back up information will not be readily available. If the venue is in the expert's own environment there may be difficulties with the knowledge engineer's travelling time and problems with interruptions which may distract the flow of the interview.

Appointment time selection

The time of day arranged for the interview influenced its success. Those which spanned lunch resulted in the elicitation of much useful information, encouraged by the informal atmosphere. However, those which occurred purely during the eating period tended to result in embarrassed parties who had difficulty in combining talking and eating.

Multiple expert situations

The danger exists, in an interview involving only two people, that the knowledge engineer will create an atmosphere of interrogation which may produce defensive rather than instructive information. This situation may be avoided with the introduction of additional participants. Where the additional party is a member of the same organisation as the expert there is little conflict as each would focus on different aspects of the overall problem. Guarded responses were noted from multiple experts who were not from the same organisation, although after a period of time, once they had the opportunity to 'size each other up', this guard would be lowered.

CONCLUSIONS

An expert's ability to offer valuable advice is not necessarily a result of knowing sufficient about a domain never to be caught out. As discussed by Swartout and Smoliar (1987), experts are able to adapt their problem solving techniques to novel or unusual situations and can often determine on the basis of the statement of a problem, whether or not it lies directly within the scope of their expertise. Experts can also explain their problem solving behaviour, which entails justifying the validity of the steps taken in reaching a solution and relating them to general principles.

It is essential, therefore, that the potential scope of the knowledge to incorporate in the knowledge base of an expert system should include understanding and experience, together with the considerations of problem solving strategy and context which occur within their circular relationship.

Specialist knowledge within an organisation represents a major asset. The training required to develop such specialist expertise can be a significant expense. However, experts can only be in one place at a time, and they are as subject to illness and changes in occupation as other humans. By capturing what is otherwise fragile and transient knowledge, its accessibility is improved and the opportunity is provided for experienced but not expert staff to use it to make recommendations and solve problems that normally require the involvement of key individuals.

Research and practical application to date has shown that expert system technology can make a significant contribution to the consistency and speed at which problems are solved.

The provision of a good user interface makes the technology very 'user friendly'. This encourages the use of expert systems by those who might otherwise refuse to employ a computer program to assist with decision making tasks. The facility to generate comprehensive reports in printed format presents benefits to both the client and the user. The detailed explanations of the advice given clarify its reliability and provide, if required, for the quick and easy amendment of the factors which influence.

Having regard to the potential of the technology, it is only a matter of time before the quantity surveying profession accept the development and use of expert systems as the way forward in undertaking the many varied management tasks now associated with their skills.

BIBLIOGRAPHY

Attarwala, F. T. and Basden, A. (1985) A methodology for constructing expert systems. *R & D Management*, **15**, 2, 141–149.

Basden, A. (1983) On the application of expert systems. *International Journal of Man-Machine Studies*, **19**, 461–77.

Basden, A. and Attarwala, F. T. (1986) *Elements of a methodology for Building Expert Systems* (publication pending).

Berry, D. C. and Broadbent, D. E. (1986) Expert systems and the man–machine interface. *Expert Systems*, **3**, 4.

Brandon, et al. (1988) *Expert Systems — The Strategic Planning of Construction Projects*. London: Surveyors Publications, RICS

Davies, M. and Hakiel, S. (1988) Knowledge harvesting: a practical guide to interviewing. *Expert Systems*, **5**, 1.

Goodall, A. (1985) *The Guide to Expert Systems*. Learned Information (Europe) Ltd, England.

Harmon, P. and King, D. (1985) *Expert Systems, Artificial Intelligence in Business*. New York: John Wiley & Sons.

Hayes-Roth, F., Waterman, D. A., Lenat, D. B. (1983) *Building Expert Systems*. Wokingham: Addison-Wesley.

Kidd, A. and Welbank, M. (1984), *Knowledge Acquisition Expert Systems: State of the Art Report 12:7*. Oxford: Pergamon Infotech Limited.

Newton, S. (1985) *Expert Systems and the Quantity Surveyor*. London: Surveyors Publications, RICS.

Olson, J. R. and Rueter, H. H. (1987) Extracting expertise from experts: methods for knowledge acquisition. *Expert Systems*, **4**, 3.

Silva, A. and Mudge, S. (1985) How an expert system was built without an expert (reported by Rory Johnston). *Expert Systems User*, November, 16–17.

Swartout, W. R. and Smoliar, S. W. (1987) On making expert systems more like experts. *Expert Systems*, **4**, 3.

Note For more information on the ELSIE expert system, please write to Imaginor System, 41 Leslie Hough Way, Salford University Business Park, Salford.

Chapter 8

Computer Aided Design

BRIAN ATKIN, *Department of Construction Management, University of Reading* and ROBERT DAVIDSON, *British Telecom*

INTRODUCTION

Traditionally, the design team has relied on paper as the primary means of communication. This has been in the form of drawings, cost plans, bills of quantities, specifications, financial statements and so forth. Each team member has tended to work in isolation, furnishing only part of the complete package, and relying on the architect or the quantity surveyor to co-ordinate project information.

Computer technology has begun to change the way design teams operate and communicate. The use of computer-aided design (CAD) techniques by designers, in particular, is becoming common and as dependency on computer-held design data increases, CAD represents a very important link in the chain of communication within and without the team.

The abbreviation CAD has two meanings: computer-aided design and computer-aided draughting. Although some designers employ both techniques, the real benefits to the client are in computer-aided design. On the other hand, the draughting side represents an internal cost/efficiency benefit for the designer.

The study undertaken by the University of Reading for the Quantity Surveyors Division of the RICS provides a good starting point for any consideration of the impact that CAD is likely to have. The study recognised the importance that CAD will play in changing and improving working methods in the construction industry. The resultant report *CAD Techniques: Opportunities for Chartered Quantity Surveyors* saw CAD as an opportunity rather than a threat. Considerable emphasis was placed on the increased involvement of the quantity surveyor at early design. Possibilities for obtaining benefits from CAD applied to the production drawings stage were also identified.

OBJECT AND METHOD

The object of this chapter is to investigate the implications for quantity surveyors arising from the increasing application of computer-aided design techniques and thereby raise the general level of awareness. Basic techniques are described and reference is made to several different geometries that are used to describe a building to a computer graphics system. This latter aspect has far reaching implications for design team members as it largely determines the extent to which data may be subsequently extracted for quantity surveying purposes. Examples of data extraction from commercially available, low cost CAD systems are given and the difficulties associated with the different approaches are discussed. Two case studies serve as useful examples of what has been achieved already in practice. As we shall see, the problems of exploiting CAD today are not caused by the technology — it provides the solutions — but by the way the industry and the professions within it operate.

DESCRIPTION OF TECHNIQUES

All CAD software is essentially graphics software and should, therefore, be expected to incorporate rapid creation, editing and manipulation of a wide variety of basic shapes (primitives). For software to be commercially successful it must perform in a way that complements and extends the skills of the designer and/or draughtsman. This applies irrespective of the geometry used to describe the building to the computer.

In some cases, a simple, two dimensional (2-D) approach may be all that is needed. The use of 2-D draughting software should be seen in the context of drawing something that has been designed already, in the expectation that productivity gains will be realised over the manual alternative. However, applying CAD to the production drawings stage limits the technology to effecting productivity gains only, as designs become frozen at this point.

Most software, once mastered, can result in manpower savings without the need to have a constant throughput of work. Much depends, however, on the degree of standardisation or repetition — preferably a combination of both — that exists within designs. The greater the degree of originality in detailing, the smaller the prospects of manpower savings. In an attempt to speed-up the rate of origination, variable shapes are normally incorporated. These are known as 'parametrics' and should be seen as essential for any software if CAD is to provide productivity gains over the manual alternative. Such features are common to all types of software and may be used successfully in conjunction with special commands or macros (user-defined routines) for assembling objects into complete designs.

Layering is a feature of 2-D draughting, with some software offering an almost limitless number of layers, each one a true transparency. In this way, it is possible to overlay services and finishes on to general arrangement drawings to help check for co-ordination of design information, and to avoid the traditional problem of layers of copy negatives obscuring the original plan. A similar capability, but using levels, can be found in some 3-D modelling systems.

Main categories of CAD software

By far the most common software is that which uses 2-D co-ordinate geometry to describe the building and its constituent parts. 2-D software is the easiest, and probably the least expensive type to develop, hence the proliferation of low cost products. Software of this type is usually described as 2-D draughting, with each drawing created as a file and equating to a different view of the building (plan, section or elevation). However, these drawing files are unconnected and any change made to one will not be automatically reflected on the others. This represents a flaw in the digital, 2-D draughting of a building that prospective users would do well to bear in mind.

Some 2-D software producers offer enhancements or extensions to their products, such as a means for contriving more complete geometric descriptions. In some cases, 3-D visualisation is possible by extruding height or depth from 2-D plans. This produces an effect that is sometimes referred to as $2\frac{1}{2}$-D, with the resultant images appearing as 3-D wire-frames. However, the latter does not offer the same degree of model wholeness or integration that is normally afforded with a true 3-D modeller. (A further form of $2\frac{1}{2}$-D can be created by a technique known as box geometry where objects are depicted by their orthogonal plan, elevation and section views.) Wire-frames are open constructions, defining objects by vertices (points) and edges (lines), and can create images that are spatially ambiguous. They should never be thought of as capable of defining surfaces. Similarly, spaces or volumes cannot be enclosed. Any perspective generation, using the technique of hidden line and surface removal, may not prove accurate as some techniques are based on dubious algorithms. Analysis of 2-D or 3-D wire-frame objects is often limited to the edges used to define the objects that are composed into the building's design, although area and volumetric measurement can sometimes be contrived.

3-D modelling offers a suitable approach to describing a three-dimensional object such as a building. 3-D models can support a greater amount of analysis because they exist in three-dimensional space. Lengths, areas and even volumes can be measured together with specifications, supporting greater horizontal integration of complementary design activities such as calculations for heat gain— heat loss and shadow pattern analysis; lighting calculations and cost estimation.

3-D modelling systems can be divided into two broad categories: surface (or boundary) models and solids (or Boolean) models. The solids approach is the closest to reality, but substantial computing is needed to support it. Sections cut through solids will always be fully defined, which helps to explain why computational overheads restrict this approach at present. Most 3-D solids modellers are used for engineering applications where relatively small numbers of solid objects are to be manipulated: solids modelling is generally unsuited to architectural design. Although all buildings are composed of solid objects, it is the form and arrangement of spaces and the specification of their surfaces and construction that are likely to be of greater interest to the architect and engineers.

3-D surface modellers are sub-divided into physical components' modellers (where the building is represented by a collection of 3-D objects) and spatial or volumetric modellers (where each space is defined by the polygonal faces of a polyhedron). Surface modellers generally allow these polygonal faces to be specified and measured. In the case of spatial modellers, it is possible to specify and measure the surfaces of enclosed spaces; define the nature of the space (whether

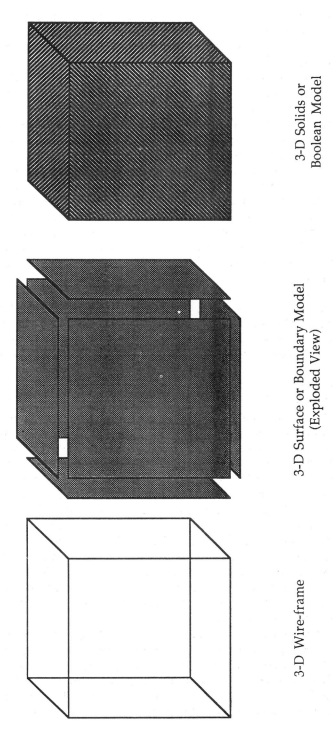

3-D Solids or
Boolean Model

3-D Surface or Boundary Model
(Exploded View)

3-D Wire-frame

Figure 8.1 3-D wire-frame, surface model and solids model

it is heated, unheated etc.); and extract other properties derived from the geometric model.

As a rule of thumb, the more simplistic the geometry that is used by the software, the more complex design details can be allowed to become. Conversely, it is possible that the closer to reality the geometry is, the less complex must be the design.

APPLICATIONS

Design, documentation and management

The design and construction process can be considered as passing through five phases which may be used as a basis for determining the areas in which CAD might be employed.

(1) Briefing: undertaking feasibility and other research-oriented studies culminating in the agreement of the requirements for the proposed building.
(2) Sketch design: generating design options that are systematically analysed, producing a single design proposal that can be developed further.
(3) Production documentation: detailed decision-making resulting in the production of drawings, specifications and bills of quantities; culminating in the acceptance of a bid.
(4) Construction supervision: monitoring and controlling the construction operations.
(5) Facilities management: monitoring and controlling the use and performance of the building under occupation.

Within each of the above, a number of applications suitable for treatment by CAD techniques can be considered.

Briefing phase

- Project appraisal
- Space requirement analysis
- Circulation analysis
- Accommodation projection

Sketch design phase
- Site planning
  - Site mapping
  - Slope analysis
  - Drainage analysis
  - Cut and fill analysis
  - View analysis
  - Site plan synthesis

- Scheme design
  - Floor plan layout generation
  - Three-dimensional spatial synthesis
- Performance analysis
  - Preliminary structural analysis
  - Shadow pattern analysis
  - Natural lighting analysis
  - Artificial lighting simulation
  - Acoustical analysis
- Presentation
  - Sketch plans, elevations and sections generation
  - Dynamic visualisation
  - Perspectives generation

Production documentation phase
- Detailed design
  - Structural member sizes and locations
  - Mechanical and electrical system design
  - Drainage system design
  - General arrangement drawings generation
  - Sectional and elevation drawings generation
  - Schedules of doors and windows generation
  - Specification generation
- Tendering
  - Bills of quantities generation
  - Construction planning
  - Bills of quantities pricing

Construction supervision phase
- Time control
- Cost control
- Production control
- Components and materials scheduling
- Valuation and payment calculation
- Management and accounting functions

Facilities management phase
- Space planning
  - Staff distribution
  - Location of equipment and furniture
- Operational management
  - Scheduling of areas, etc. for cleaning and similar support services
  - Location of security and protection systems
- Maintenance management
  - Planned preventive maintenance scheduling
  - Job costing

MAJOR ISSUES

Which geometry?

For certain applications, such as refurbishment, domestic scale building work and facilities planning, the 2-D approach may be adequate, as only a relatively small number of orthogonal views of the building may be needed. For modelling most building designs, the 2-D approach may be cumbersome and dimensionally inadequate. 2-D draughting systems should, however, be used to produce orthogonal views for documenting a detailed design, but they should take their basic geometry from 3-D models. It is, therefore, not a choice between 2-D draughting or 3-D modelling software, but a matter of deciding which 3-D modelling product to use for early design and which 2-D draughting product to use for later production drawings work.

Perhaps the greatest opportunity that 3-D modelling software presents is the ability to generate larger numbers of potential design solutions than are possible manually. If this can be matched by appropriate database support, containing cost and energy data, iterations of design generation and analysis can become quite rapid — down to less than one hour.

For this reason, it is likely that the 3-D modelling approach will soon become an established tool to aid early design. Links to other software, such as spreadsheets, word processors and relational databases are possible now, encouraging greater horizontal integration of design activities. However, achievements will be minimal if the 2-D draughting approach is used exclusively to describe building designs.

What are the target areas for quantity surveyors?

It is impossible to talk about CAD as though it were a homogeneous topic. From the quantity surveyor's viewpoint, working with a variety of organisations representing different professional interests means that he/she will encounter any or all of the types of software discussed earlier. The problem is heightened by the presence of different operating systems and by the physical dislocation of the parties involved in a typical project. How then does the quantity surveyor make a start?

Areas worthy of consideration, although not in any order of priority are:

- cost modelling;
- bills of quantities production;
- data management.

Cost modelling
An area of considerable potential for the quantity surveyor is during early design. It is here that the 3-D CAD systems discussed earlier would come into their own.

Presently, direct communication with a client about his requirements and the cost implications of options is inhibited by the absence of a system to illustrate rapidly the form and nature of potential design solutions. Traditionally, identification of a possible design solution requires the production of a sketch plan and its subsequent costing. This introduces a delay into the process and removes the possibility of developing ideas interactively with the client. By using a CAD system, the quantity surveyor can work closely with his client to refine options and assist in evaluating design solutions. Additional software which produces development budgets would allow the client to see the total effect of the decisions he is making.

Key requirements are the need for effective on-line databases and links between the CAD system of the designer (or design team) and the computer systems of the quantity surveyor. Speed of response will be a deciding factor in its successful operation. There is little value in the designer having the facility to generate several options rapidly using CAD, if they cannot be costed equally rapidly.

Compatibility of operating systems raises other issues of a more practical nature. Will the interface be through the medium of a PC, a workstation or a minicomputer? If the answer to this question can be found, with which proprietary systems will the quantity surveyor need to communicate? Fortunately, tools exist for exchanging data between systems, that is, between specific pairs of systems and generally between different systems using a common format (for example, DXF, IGES and the forthcoming system, STEP). But this can be at the expense of loss of precision and of considerable processing time.

Bills of quantities production
A bill of quantities can rightly be thought of as simply a type of cost model. However, the notion of using a CAD system to produce bills of quantities raises other issues which are best considered separately. For instance, the task of measuring from a CAD produced drawing has an air of duplication (and indeed unreality) about it. This is especially true when one considers that the data probably already exist within the computer which generated the drawing in the first place! Unfortunately this hypothesis is easier stated than proven.

It is a fact that CAD software can now provide certain quantities, but the industry is a long way yet from pressing the button that generates full bills of quantities.

What does exist falls into two categories. The first is probably best labelled as 'scheduling' where the software can list out component type data, for example, sanitary fittings, doors and windows captured as part of the CAD process which itself called on these items from preselected design data or libraries created for the project in hand. Beyond this 'listing' function one can find certain measurement features built into most software packages which can provide lengths, areas and volumes of defined objects.

The second category is purpose written software which sets out to provide quantities in a specified area of work. Probably the best example of this approach is the work by Monk Dunstone Associates on Mobil service stations (discussed in a case study at the end of this chapter), enabling the production of the quantities. This can only operate in a defined area, for example, a 'standard' type solution or at least one where there is some level of consistency in the data being handled. In this latter area, the PSA have commissioned ABS Oldacres to look at the potential for automatic generation of quantities and associated data from CAD in the mechanical and electrical services field.

Bespoke software preparation is an expensive operation, especially in an area where changes in standards, methods of measurement, etc. are the norm. So, it seems likely that the long term solution to bills of quantities production must come from competition amongst software vendors to provide better 'add on' packages to their basic software. Clearly, the simpler the method of measurement the faster this will come about.

Data management
A third possibility which CAD creates is that of managing data for projects in a more effective way. This falls into two types: a graphical database and a non-graphical database. The first provides the graphical description of the project as it evolves. The second is closer to the quantity surveyor's interests and responsibilities in the area of specification data, component attributes, quantities, costs, etc. The quantity surveyor would be well placed to manage these data on behalf of the project team and also to organise 'project profiles' once a project is complete. These latter data would augment the traditional cost analysis and enable the project team to search for and retrieve information on tried and tested solutions.

EXAMPLE METHODS OF DATA TRANSFER

ASCII file

Compatibility between different computer systems can be achieved by transferring data files in ASCII format. Most software, irrespective of the computer being used, can output data in this format. For instance, the majority of 3-D CAD and engineering design systems have the ability to produce quantified reports that can be defined by the designer. These reports may include information which is required for further calculations either by the designer or by another consultant such as the quantity surveyor. Printed reports are generally used by the designer to check areas or schedule components. However, instead of printing, the information may be written to computer file, and this will almost always be in ASCII format. Information can, therefore, be transferred to different software on the same computer or to another computer in any location using a communication link or magnetic media in tape or floppy disk format: software is also required to take account of different operating systems. This information could then be used by the quantity surveyor for estimating purposes or, for instance, by the services engineer for calculating heating and cooling requirements.

It is very easy to edit ASCII data files using a text editor or word processing software. This does mean, however, that data could be deleted, but with careful use the layout and content of the data extracted from the CAD system can be enhanced or altered, enabling other software to import the data for further manipulation and processing.

```
1nr   CDORO39    :CCDA?        (    )    800 x 1600 x 2050

                            : Clear varnish to BS1234
                            : 6mm georgian wire
                            : Full glazed
                            : Not rated
                            : Undecided

2nr   UIRN001                   (32   )

1nr   CDORO39    :CCDB?        (    )    800 x 1600 x 2050

                            : Clear varnish to BS1234
                            : 6mm georgian wire
                            : Full glazed
                            : Half hour
                            : Undecided

2nr   UIRN001                   (32   )

4nr   CDORO40    :BADBB        (    )    900 x 955 x 2050

                            : Intumescent paint to BS1234
                            : 6mm sheet
                            : Full glazed
                            : Half hour
                            : RC lintol

   1. COST                    : 150 POUNDS

1nr   CDORO46    :CCDAB        (    )    900 x 955 x 2050

                            : Clear varnish to BS1234
                            : 6mm georgian wire
                            : Full glazed
                            : Not rated
                            : RC lintol

   1. COST                    : 120 POUNDS

2nr   UIRN001                   (32   )

1nr   CDORO47    :CCDAB        (    )    900 x 955 x 2050

                            : Clear varnish to BS1234
                            : 6mm georgian wire
                            : Full glazed
                            : Not rated
                            : RC lintol

   1. COST                    : 120 POUNDS

2nr   UIRN001                   (32   )
```

Figure 8.2 Example 'print-out' of specification data for subsequent manipulation by word processor

3-D model to spreadsheet

Several 3-D modelling software packages are now available and in use within design offices. It is impossible to say which of these packages is best as many factors must be taken into account. Clearly, an appropriate package needs to be selected here for illustrative purposes. To this end Personal Architect produced by Computer Vision for use on IBM PCs or compatibles, is one such package. As far as spreadsheets are concerned, there are probably dozens in use. Lotus 1−2−3 is, however, the most well-known and well-used spreadsheet.

The Architectural Design Module of Personal Architect can be used to create building designs which may then be costed rapidly. Reports can be designed to output quantity information in composite item format within elements. Such reports can then be converted to ASCII file format and edited. For instance, data may be transferred to a pre-prepared Lotus 1−2−3 spreadsheet containing the rates for the composite items. Items are extended rapidly and a total cost estimate can be produced in seconds.

3-D model to quantity surveying systems

Personal Architect can also be used to generate the design and then produce data in ASCII file format. Data can then be edited using a word processor to delete unwanted space and headings, and also to add additional data such as sort codes and item description references. The word processor can also be used to duplicate data records where more than one measurement item is needed for a single data record output by the CAD system. For example, 'net room perimeter' could be used to generate measurement items for skirting, ground and painting.

A quantity surveying computer system, such as CATO (Computer-Aided Taking Off) can be used to process data extracted from Personal Architect. Within the CATO software portfolio is a program called CATO Link. This software has been designed specifically to read ASCII data into the main CATO measurement programs. One of the primary reasons for its introduction was to allow data, from different CAD and engineering design systems, to be passed semi-automatically to CATO in the future. Its main use currently is for passing data generated by spreadsheets and databases into CATO.

Before ASCII data can be read by CATO, the user must first state the exact format of the data record. This must include the start position, format and/or length of:

- descriptive text generated by the CAD system for annotation purposes;
- item description references added by the quantity surveyor;
- sort codes added by the quantity surveyor;
- the actual quantity to be recorded.

CATO Link can read every line of the ASCII data file and generate a measurement record for each line or optionally can presort all like items using up to nine

---

ROOM FINISH QUANTITIES
(BY ROOM & CELL)

PERSONAL ARCHITECT
ARCHITECTURAL DESIGN MODULE
VERSION 2.0

FINISH QUANTITIES — FLOOR & CEILING

ROOM NAME   : AIR

Volume Type	:	1	Floor Area- net	:	0.83 m²
Room Perimeter- total	:	3.70 ml	Floor Opening Area	:	0 m²
Room Perimeter- net	:	2.79 ml	Interior Heated Area	:	0.83 m²
Ext. Threshold Length	:	0 ml	Interior Unheated Area	:	0 m²
			Exterior Terrace Area	:	0 m²
Volume (Cu. Ft.)	:	2.06 m³	Crawl/Mechanical Area	:	0 m²

ROOM NAME   : BATHROOM

Volume Type	:	1	Floor Area- net	:	6.68 m²
Room Perimeter- total	:	13.70 ml	Floor Opening Area	:	0 m²
Room Perimeter- net	:	12.79 ml	Interior Heated Area	:	6.68 m²
Ext. Threshold Length	:	0 ml	Interior Unheated Area	:	0 m²
			Exterior Terrace Area	:	0 m²
Volume (Cu. Ft.)	:	16.63 m³	Crawl/Mechanical Area	:	0 m²

ROOM NAME   : BED 1

Volume Type	:	1	Floor Area- net	:	11.01 m²
Room Perimeter- total	:	13.30 ml	Floor Opening Area	:	0 m²
Room Perimeter- net	:	11.48 ml	Interior Heated Area	:	11.01 m²
Ext. Threshold Length	:	0 ml	Interior Unheated Area	:	0 m²
			Exterior Terrace Area	:	0 m²
Volume (Cu. Ft.)	:	27.41 m³	Crawl/Mechanical Area	:	0 m²

Figure 8.3 Example 'print-out' from 'Personal Architect' software showing volumetric, surfaces and perimeter quantities — summaries of these data can be easily imported into a spreadsheet for post-processing.

different sort criteria and then create a single measurement record for each total. If presorting is required, the quantity surveyor has to enter the sort name, start position and length of each of the sort fields required.

CATO Link can, if required, read a file more than once in order to read from records which have a number of quantities in them. Item description references relating to the CATO Approximate Estimating Library are added to the CAD data file and read into CATO line by line. A bill of approximate quantities is then produced and priced. Prices are held against each item description in the Library and updated in line with changes in market conditions, from the date the rates are established to the anticipated tender date for the project. The Library is structured in elements and as data are read from the CAD file, element codes are added automatically and then passed into the priced bill file. From the priced bill, an elemental cost analysis is produced into which may be added amounts for services, site works, drainage, preliminaries and contingencies.

The time taken to process these data and to generate the elemental cost analysis is only a matter of minutes. More time is, however, needed to edit and enhance the information within the ASCII data file, and creating and pricing descriptions not already held in the CATO Approximate Estimating Library. Nevertheless, the overall time taken is small compared to that which would have been taken if the quantity surveyor had to measure manually all the quantities otherwise generated by the CAD system.

2-D draughting to quantity surveying systems

AutoCAD, although not necessarily the best CAD system, is probably the most widely installed worldwide. The actual use of AutoCAD by designers is confined mainly to production drawings work, but it can have an application much earlier. AutoCAD draughting and design package and the AutoCAD AEC (Architecture Engineering Construction) package can be used to generate alternative outline designs for subsequent analysis. Designs are drawn as wire-frames so that alternative shapes can be input readily. As much detail as possible for the elements then needs to be incorporated into the wire-frames, so as to give the quantity surveyor the maximum amount of measurement data for the different shapes.

Measurement data can then be gathered by 'pointing' to areas etc. on the screen and then passing the quantities produced into a data file. Data files are produced for each of the design alternatives and include as much element quantity measurement information as could be extracted from the wire-frames. This is done using the special programming language, AutoLISP in the AutoCAD AEC package.

All data files are in ASCII format and can be read by CATO Link (see above) to generate an elemental cost analysis. The data files can also be read by a preprogrammed Lotus 1–2–3 spreadsheet to calculate a cost plan.

Electronic data exchange

Clearly, data can be exchanged between otherwise incompatible computer systems by reducing the data to the lowest common denominator, that is, ASCII file format. More elegant ways of transferring data are needed at an industry-wide level if the benefits of integrated CAD are to be realised. The co-ordinated approach being adopted by the project working groups within EDICON (Electronic Data Interchange for Construction) aims to ensure that the 'message structures' for the electronic interchange of design, specification, product data and bills of quantities information respect and reflect their different and common needs. If full, economic collaboration is to take place between CAD users and quantity surveyors it is vital that the correct message structures are established from the start. Electronic data interchange is the logical next step in the long drawn out 'coming-together' of CAD and quantity surveying systems.

CASE STUDY: LLOYDS OF LONDON

The measurement techniques adopted for the mechanical and electrical services of the Lloyds of London building illustrate an interesting bridge between early manual techniques and the automatic generation of measurement and descriptions.

The Lloyds project comprised approximately one hundred and forty subcontract packages, many of which were quite small. Bills of quantities were, however, prepared for a significant number of major packages. The techniques involved are well established and documented and it was using these techniques that the majority of the bills of quantities were processed.

Under normal circumstances these techniques would have been used for the production of the tender documentation. Unfortunately, changes in Lloyds requirements, ironically involving the projected growth in the use of computers in the new building, meant that elements of the services installations had to undergo major redesign. Consequently, information was not available in sufficient time to enable normal taking-off and billing techniques to take place if sacrosanct programme dates were to be safeguarded.

In order to circumvent, what, at the time, was a major difficulty, a series of programs was written to permit the 'blind' measurement of the pipework and ductwork elements. Detailed discussions were held between Monk Dunstone Associates and Ove Arup and Partners, the consulting engineers, resulting in the drawings being prepared with coded information to denote type of material, size, insulation, over-cladding, etc. At this stage, whilst routes and other important information were known, sizes could not be calculated nor could materials be specified. Codes, together with dimensions for lengths of pipes or ducts and their fittings, were entered directly by the 'taker-off' at a workstation. As soon as the engineers were able to size and specify their requirements the information was allocated to the relevant codes and fed into the computer to permit the sorting and successful production of full bills of quantities.

The relatively discrete number of items/materials involved in the services installations (for instance, pipes, ducts, insulation, fixings and cable) enabled requests for the evaluation of potential variations to be measured as 'mini' bills of quantities. These variations were then priced using the subcontractor's rates which were stored against the subcontract bill items. Once adopted and issued under an architect's instruction, the measured variations were added to the appropriate section of the database so that an up-to-date final account total for each of the services' subcontracts was always available. This process aided the reconciliation of interim valuation applications and the speedy negotiation and settlement of the services' final accounts.

CASE STUDY: MOBIL OIL COMPANY

In the early 1980s, the Mobil Oil Company decided to acquire a CAD system to standardise and improve the presentation of their drawings for their retail sales outlets. In addition, and in a quite revolutionary way, they wanted a system that would assist in the automatic generation of bills of quantities. The CAD system selected was GDS (General Draughting System) marketed by Applied Research of Cambridge (ARC) who have subsequently become part of the McDonnell Douglas Group of Companies. ARC appointed Monk Dunstone Associates and ABS Oldacres Computers to assist in developing a measurement link.

It must be emphasised that the Mobil service stations use standard components throughout the country so that, for example, a petrol pump island is one of only two designs, wherever it is located. Although the total number of components is quite large, it is absolutely finite. The integration of CAD and bills of quantities systems was aimed not only at the standardisation of documentation but also at post-contract control.

The development uses the GDS library as its base. Each item on the service station site (standard building types, pump islands, paving, service and drain lines and trenches, storage tanks, etc.) is drawn and held as a GDS library item with a quantity of unity, as number, linear, square or cubic. For each of these objects full quantities were measured to comply with the SMM including, where appropriate, adjustment dimensions. These adjustments ensure that, for example, a fuel pump island contains not only the dimensions for the island but also the quantities to deduct the forecourt paving for the footprint area it occupies. For areas such as paving, the CAD system generates an area which is applied to all the descriptions contained in the single square metre held as the library item.

On completion of the layout drawing by Mobil's design engineer, the GDS system generates the drawings. Is is then possible to process full bills of quantities within a couple of hours. The Mobil engineer is only required to consider items for site clearance, demolition, bulk cut and fill (if applicable) and to edit, using a word processor, standard preliminaries and preambles.

The benefit to Mobil of this scheme is quite dramatic. They can have a printed bills of quantities to send to tenderers within a day of the completion of design drawings and have made substantial savings in professional fees.

From time to time, the library of standard objects is updated as new components are required or as existing components change. The appropriate additional quantities are measured and added to the database of dimensions.

CONCLUSIONS

This chapter has highlighted some of the implications and opportunities that CAD holds for the quantity surveyor. What is clear is that a growing number of designs are being produced with the aid of a CAD system. This trend will inevitably influence the quantity surveyor's working methods.

Perhaps the most likely scenario for the future is that of the quantity surveyor with his own CAD workstation communicating with the rest of the design team. As the design evolves, cost advice can be provided rapidly and the data to be extracted subsequently for contract documentation can be systematically introduced and organised. Before that stage is reached many problems need to be overcome, not least the ones concerning the variety of equipment, operating systems and software tools with which the quantity surveyor might need to communicate. A common, industry-wide approach to electronic data interchange would be more than welcomed.

ACKNOWLEDGEMENTS

This chapter is based largely on a select number of papers, presented at the Barbican in November 1988, by the following people whose contribution is gratefully acknowledged: Geoffrey Ashworth, Chris Dinnewell, Brian Edgill, Martin Hawkins, Geoff Hawkings and Brendan Patchell.

BIBLIOGRAPHY

Atkin, B. L. (1987) *CAD Techniques: Opportunities for Chartered Quantity Surveyors.* London: Surveyors Publications, RICS.

Day, A., Faulkner, A. and Happold, E. (1986) *Communications and Computers in the Building Industry.* Cambridge: Construction Industry Computing Association.

Hamilton, I. and Howard, R. (1984) *Prescribing CAD Systems for use in Health Building.* Cambridge: Construction Industry Computing Association.

Hamilton, I. and Winterkorn, E. (1985) *CAD Systems Evaluated for Construction.* Cambridge: Construction Industry Computing Association.

Mitchell, W. J. (1977) *Computer-aided Architectural Design.* New York: Van Nostrand Reinhold.

Mitchell, W. J. (1984) CADD applications in architecture, engineering and construction. *CAD-CAM Handbook*, ed. Teicholz, E. New York: McGraw Hill.

Port, S. (1984) *Computer Aided Design for Construction.* Oxford: Blackwell Scientific Publications.

Port, S. (1989) *The Management of CAD for Construction.* Oxford: Blackwell Scientific Publications.

Wager, D. and Wilson, R. (1986) *CAD Systems and the Quantity Surveyor.* Cambridge: Construction Industry Computing Association.

Wix, J. and McLelland, C. (1986) *Data Exchange between Computer Systems in the Construction Industry.* Bracknell: Building Research and Information Association.

Chapter 9

Integrated Databases

BRIAN EDGILL*, *Property Services Agency, Croydon* and JOHN
KIRKHAM, *Information Technology Institute, University of Salford*

THE EMERGENCE OF THE DATABASE

The growing use of information technology (IT) in the construction industry has
made quantity surveyors increasingly aware of the volume and potential value of
the data which they routinely handle as part of their day-to-day work. Information is
the life-blood of the construction industry and the quantity surveyor occupies a
key position in the total design/construction process. By applying computers to a
growing proportion of quantity surveying activity, the quantity surveyor is placing
himself in a unique position to capture, analyse, store and re-use the data which
such procedures generate. In the past, IT's main role was simply to speed up the
manual process. The advent of database technology enables IT to be more
effectively included in the quantity surveyor's kit of tools.

The quantity surveyor has always been at the forefront of computer applications
from the mid-60s when computers were applied to the obvious data intensive task
of bills of quantities preparation. Since then the application areas have extended
to cover the full range of activities from early cost advice to financial settlement.

This chapter addresses the potential of organising the data into an integrated
database for the quantity surveyor's office. As the nature, size and complexity of
the data will vary from office to office so will the choice of data handling
techniques. These could range from straightforward off-the-shelf packages, through
more sophisticated (but nevertheless purchasable) packages through to the possibility of creating one's own bespoke database.

To provide guidance to such a varied audience is not easy. The approach has
therefore been to concentrate on the principles involved and to explain the basic
techniques of data organisation. This then will provide the basic understanding
needed whatever level of application is pursued. The examples of data models
provided have been drawn from case studies made as part of a study (Brandon
and Kirkham 1989) of integrated databases for quantity surveyors undertaken for
the Royal Institution of Chartered Surveyors (RICS) by the University of Salford.

---

* The views expressed are those of the author and not necessarily those of the Property
Services Agency.

DATABASES

The quantity surveyor is probably the most active data handler in the design/ construction process. The bill of quantities itself represents a vast amount of data but that is clearly only part of the story. The other documents — financial appraisals, estimates, cost plans, final accounts, etc. — are basic to the quantity surveyor's role and to the overall process. Whilst key to the efficient translation of client needs to completed buildings and structures they also can provide a source of valuable historic data for use on future projects. These then, in IT form, represent the potential of the quantity surveyor's database.

The term database has been used loosely since the mid-1960s to indicate any large integrated collection of data found on a computer. In some circles it is still used in this sense and this can cause some confusion. A more precise definition which describes the concept as used in this report is 'a generalised integrated collection of data which is structured on natural data relationships so that it provides all necessary access paths to each unit of data in order to fulfill the differing needs of all users' (Deen 1977).

The value of databases is that users can find information more easily, more efficiently and more consistently. In the more advanced forms the relationships between data are clearly expressed so that the implications of a change in one item will be reflected in those other items which are associated with it.

The integration of the data helps to ensure that information does not become out of date because, for example, messages are not passed on or the repercussions have not been fully realised. It also ensures that redundant information is taken out of the system. In manual systems the problems of communication and the physical chore of keeping files up to date very often results in faults of this kind.

The concept of 'natural data relationships' allows the association between items to be expressed and cross-referenced in order to aid such matters as the speed of retrieval of information or the rapid updating of information when one item is dependent on another.

Finally, it is important that a wide variety of users of a system can interrogate and manipulate the information in the database in an efficient and effective manner. This requires a good user interface and a thorough understanding of how the system will be employed.

Quantity surveyors are not primarily engaged in 'production line' work and each contract undertaken is always different in some aspect from the previous one. Consequently the development of an integrated computer database for quantity surveyors is formidable because of the difficulty of determining and documenting the complex relationships between the data which has to be stored. For example for each contract there are different relationships between members of the firm and outside organisations such as architects, local authorities, contractors, engineers, government, etc. There are also intricate, and changing, relationships between all the documents that are used in a project and the people and organisations that use them.

Many computer applications in the quantity surveying area merely automate such manual activities as payroll, time management and the preparation of the bill of quantities. It is important that quantity surveying organisations realise that these computer systems should do much more than this. Computer systems could and should improve the way quantity surveying organisations do business, changing their relationships with clients, designers and contractors. Interaction with computer systems could also change the way decisions are made, alter the structure of the organisation and allow the creation of new business and methods of working.

The integrated computer database required for a quantity surveyor is very complex. To build such a system requires the adoption of a systematic and rigorous approach or method. The method described here is the information engineering method (Martin 1987) and several of the diagrams are drawn from or based on that method. The use of this method does not result in one single computer system but a number of computer systems that react together in an integrated manner. The integration is achieved by the development of an integrated database which holds all the data pertinent to the organisation. The computer procedures then access the database to obtain the relevant information for the quantity surveyor.

INFORMATION ENGINEERING

The amount of detail information required within an organisation depends on the level at which the individual works. Senior management are concerned with the global view, summaries of data, whereas takers-off are concerned with minutiae. It would be foolish to build a computer system for the taker-off without considering the needs of senior management. Unfortunately the amount of effort required to build an integrated computer system to satisfy all the operations carried out would be unlikely to be cost-effective (see Figure 9.1) because of the overwhelming amount of detail that would have to be collected.

A method must be developed which allows identification of the 'best' parts of the business to be computerised. This is commonly called the 'divide and conquer' strategy. An overall plan for the enterprise is formulated, part of this plan is analysed in detail, part of the analysis is used to design a computer system, and finally part of the design is constructed to form a computer system. These four stages are shown in Figure 9.2. The first two stages are independent of the technology as they are logical models of the required system. The final two stages are dependent on the hardware and software, the target environment, on which the system is to be implemented. Because the system is being constructed according to an overall strategy all the computer systems can be independently constructed yet will form part of an integrated system.

These four stages must be developed in conjunction with the user, in this case the quantity surveyor. Traditionally the detailed requirements for computer systems comprised solely English prose. This was not very successful as the prose was verbose and prone to misinterpretation. The current approach is to build data and process models which show *what* the business does but not *how* it does it. The data model describes what data is required and the process model describes what processes operate on that data and the interrelationships between processes. The

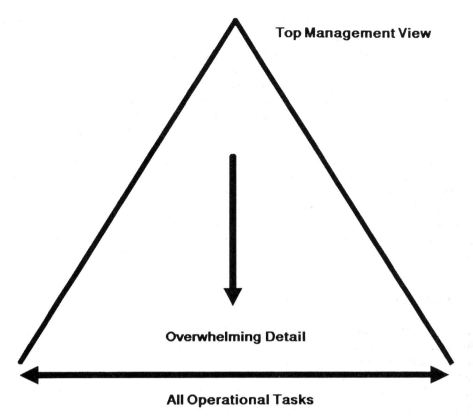

Figure 9.1 Levels of detail

two models are complementary. The process model can be used to check that all the data have been defined and the data model to check that no processes are missing. The models are usually presented as pictures which can be easily understood by users. These models are always independent of the technology employed and can be used to implement the system on any suitable computer. The four stages are described below:

Plan Concerned with top management goals and critical success factors and how technology can be used to create new opportunities and give competitive advantage. A high level overview is created showing the processes and data needs.

Analysis Concerned primarily with understanding what data and processes are needed for this particular area, how the processes and data items interrelate.

Design Concerned with the detail of how selected processes work in this area.

Construction Implementation of processes on the computer system using fourth generation languages or code generators.

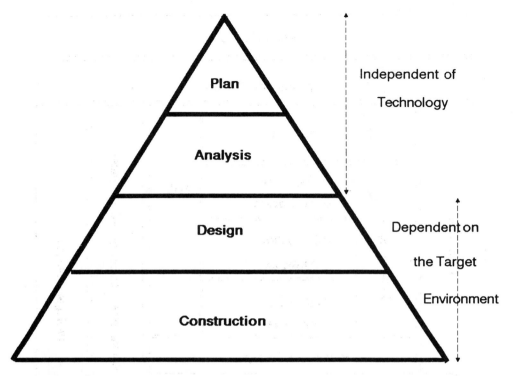

Figure 9.2 Divide and conquer

BUILDING BLOCKS OF INFORMATION ENGINEERING

Information engineering is not a rigid method which forces the development of computer systems down a rigid path. It recognises that it is not desirable to 're-invent the wheel' and that there may be suitable computer systems already available or that one can be created using a simple spreadsheet. Figure 9.3 shows the different building blocks used in information engineering and how these can be used to implement operational systems in a variety of ways once the initial planning is complete. A brief explanation of the various blocks follows.

Enterprise model and strategic information planning

The objectives of the enterprise and its components are established and the information needed to accomplish its objectives is determined.

Information analysis

This block creates an overview of the data needed to run the enterprise. The types of data, called entities, that must be held and the relationships between them are represented.

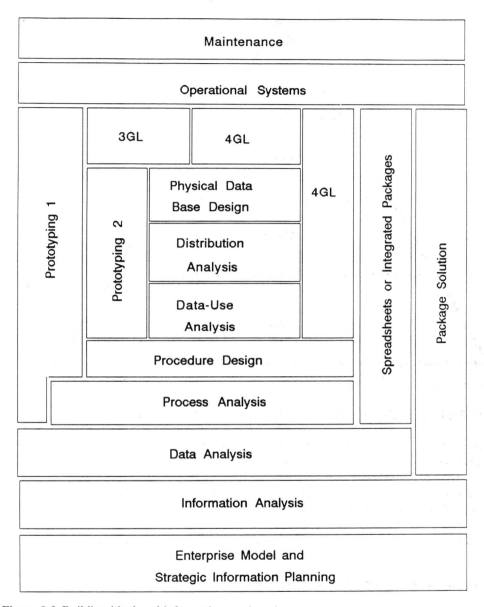

Figure 9.3 Building blocks of information engineering

Package solutions

The required computer system may already exist as a package. This package could then be purchased and put to work. As the information analysis has been completed the package should fit in with the rest of the systems to be developed.

Data analysis

Data analysis describes the types of data required. Data analysis creates the details of the data necessary to implement the database system.

Spreadsheets or integrated packages

The data analysis has been carried out and the detailed data model built. It may then be possible, if the model is a relatively simple one, to develop the required system using a spreadsheet or integrated package on a personal computer.

Process analysis

A particular area is selected for analysis and the business processes are analysed in data and correlated with the data model built previously.

Procedure design

The business processes are converted into computer procedures and combined to build up the computer system.

Fourth generation language (4GL)

The database files and procedures can now be rapidly implemented using a fourth generation language such as ORACLE, FOCUS, PICK, dBase III, etc.

Prototyping

Fourth generation languages can be used to build prototypes so that the user can see what the finished system will look like. Two approaches are possible. In the first, prototyping 1, these prototypes will ultimately become the working system. The prototype can be constructed using either data analysis or process analysis as the starting point. In the second, prototyping 2, these prototypes can then be used as a design document to build the system using a third generation language, for example COBOL, BASIC, PASCAL, etc. Typically third generation languages enable the building of computers systems which are more efficient than fourth generation languages in the use of computer resources. This difference is becoming less marked. Fourth generation languages usually result in much lower maintenance costs as fewer statements are required to carry out a particular function. They also help compress the time in which applications can be effectively produced.

Performance analysis

Data analysis results in a logical design for the database. Before the database can be implemented an analysis is made of its size, usage and complexity.

Distribution analysis

Parts of the database may have to be physically located at different sites within the organisation. An analysis is made to determine the optimal way that this distribution should be accomplished.

Physical database design

If the computer system has to be designed to have a rapid response or many simultaneous users, for example a large construction project, then a detailed analysis will be carried out to determine the optimum design on the particular computer system selected. The system can then be implemented using a third or fourth generation language.

Operational systems

Once the system is developed it is put into operation.

Maintenance

Maintenance, in computing terms, means altering the system because of changes in the business or correcting mistakes in the design. If the system has been designed and constructed in a methodical way it should require little maintenance.

INFORMATION ANALYSIS

A more detailed discussion of how to build a data model for the typical quantity surveyor practice will now be developed.

Entity

The parts which comprise a data model are things which exist and have certain properties (facts) and which may be related in some way to other things. In data modelling the thing is an entity, about which we want to store information, and the facts about the thing are called attributes. Entities may be objects such as an employee, client, part of a building, or construction site, or events such as starting the construction of a building, or activities such as the construction of a project. Each entity will have certain properties (attributes) which in the case of an employee might be name, employee number, age, sex, salary, bonus, etc. Each attribute may or may not have an attribute value. It is not necessary that all attributes have a value at a particular instance, only that it should be possible to list the properties which could be useful at some time.

It is important to distinguish between an entity type and the instance of an entity. The entity employee is a type and J. Doe and A. N. Other are instances of the same entity type employee. If K. Smith had just joined the company and had not been awarded a bonus he would still belong to the employee entity type even though the attribute value of his bonus was missing or null.

Relationship

In the 'real world' there are relationships between things. These must be represented within our data model and it is possible to distinguish between a relationship type and a relationship instance. A relationship type is a statement of the concept that one entity type may participate in a relationship with another, or the same, entity type.

Some relationship types may be treated as entity types with attributes. However it is the existence of the relationship which is important together with the properties of the participating entities. For example 'employee works on project' has the relationship 'works on' between employee and project and it is only the relationship that is of interest. On the other hand 'client starts construction of project' seems as though there is a relationship 'starts construction of' between client and project. However this relationship may have attributes of start date, authorisation, etc. and should be an entity type in its own right which has separate relationships with client and project. Relationship types do, however, have four simple properties of name, cardinality, optionality and role.

Name

The relationship can be named in two directions. An active verb being used in one direction and a passive one in the other. For example the active verb has the form 'employee works on project' and the passive form of 'project worked on by employee'.

Cardinality

The cardinality of a relationship is the number of instances of each of the participating entity types which partake in a single instance of the relationship type. The basic choice for each of the entity types is to whether an instance of the relationship will involve one instance of the entity type or many instances. In the example of employee works on project the relationship is said to be many-to-one (usually labelled N:1) between employee and project as many employees will be working on the project. Conversely the relationship between project and employee is one-to-many (1:N) as one project will have many employees working on it. If, however, employees can work on different projects then the relationship will be many-to-many (M:N).

Optionality

The participation of entities in relationships can also be optional or mandatory. Before construction starts on the project there may be no employees assigned to it. Thus the entity type project may not have workers assigned to it and its participation in the relationship type 'worked on by' is optional. Similarly the entity type employee can exist without working on a particular project and its participation in the relationship 'works on' is optional.

Roles

The participating entities can play different roles. The relationship 'works on' between employee and project did not tell us that Joe was the quantity surveyor on the project. If we introduce a new relationship 'quantity surveyor on' then:

Joe	works-on	Forth-Bridge
Joe	quantity-surveyor-on	Forth-Bridge
Alex	works-on	Severn-Bridge
Fred	works-on	Forth-Bridge
Mary	works-on	Forth-Bridge
Joe	works-on	Severn-Bridge
Joe	quantity-surveyor-on	Severn-Bridge
Ted	works-on	Severn-Bridge

There are two relationship types between employee and project but both have significantly different roles.

Entity—relationship diagrams

A convenient shorthand notation for a data model is an entity—relationship diagram. This is a graphical representation of the elements of the data model namely entities, relationships, optionality and cardinality. An entity is represented by the name of the entity in a rectangular box, see Figure 9.4.

The default relationship between entities is a mandatory one-to-one (1:1). This

Figure 9.4 Entity types

is drawn as a line between the entity type boxes with a bar, signifying the 1:1, across the line at each end. The relationship is named by placing the name next to the entity from which to read. For example Figure 9.5(a) would read from left to right as 'Employee works on always one project' and from right to left as 'Project worked on by always one employee'. Notice the word mandatory has been replaced by the word always to make the reading of the relationship more natural and 'English like'.

An example of the data and the relationships which would be stored on a working database are shown in Figure 9.5(b). The relationship has only been shown in one direction. In this case the relationship 'works on' should be read in the direction of the arrow. The other relation 'worked on by' has been omitted for clarity.

(a)

(b)

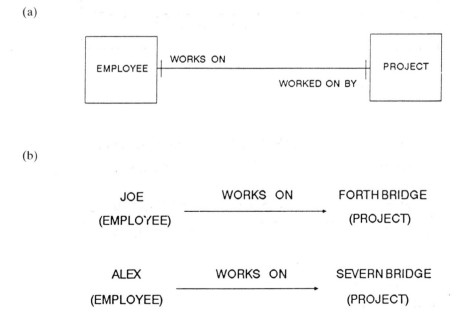

Figure 9.5 Mandatory 1:1 relationship

The optional relationship between entities is an optional one-to-one (1:1). This is drawn as a line between the entity type boxes with bars (1:1) and the circles (o for Optional), across the line at one end. For example Figure 9.6(a) would read as 'Employee works on sometimes one project'. Notice that the word optional has been replaced by the word sometimes. The passive relationship has not been changed.

The example in Figures 9.5(a) and (b) would not be very useful in practice as all employees would have to immediately start working on a project. Figure 9.6(b) removes the restriction that all employees must work on a project. The active relationship in Figure 9.7 reads as 'Employee works on sometimes one or more project'. The word sometimes has replaced the word optional and is represented by a small circle and the one or more (many) is represented by a 'crow's foot'. Note that 'sometimes one or more' in fact means 'zero, one or many' as the entity has the option of participating in the relationship. The passive relationship has not been changed.

The active relationship in Figure 9.8(a) reads as 'Employee works on always one or more project'. The passive relationship has not been changed. Figure 9.8(b) shows how each employee must work on more than one project.

(a)

(b)

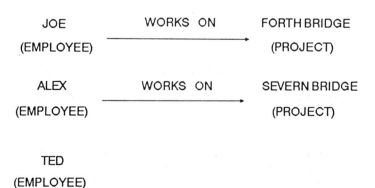

Figure 9.6 Optional 1:1 relationship

Figure 9.7 Optional 1:M relationship

(a)

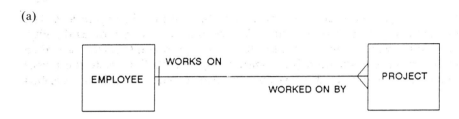

(b)

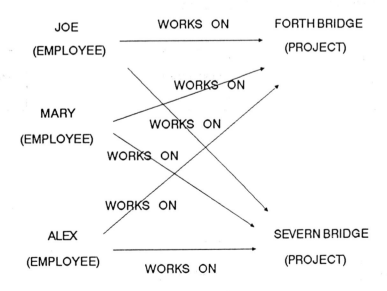

Figure 9.8 Mandatory 1:M relationship

Entity—relationship approach

The entity—relationship approach to building the data model is 'top-down' as it attempts to first identify the entity types and then the relationships between them. To demonstrate the approach a simple quantity surveying practice model will be described.

'A client commissions a project or projects and all projects must have a client. The project has employees working on the project and for management reasons employees only work on one project at a time. All employees are given a different grade such as, partner, associate, etc. The employees have to account for the time spent and fill out time sheets.'

There are five object type entities which have been drawn in Figure 9.9. The singular of the entity is used as any pluralities are taken care of by the cardinality. The relationship types (bold), between the entity types (underlined), are then determined, drawn on a diagram (see Figure 9.10) and labelled. Note the names are not quite the same as in the text but conform to the conventions described earlier.

'A client **commissions** a project or projects and all projects must have a client. The project has employees **working on** the project and for management reasons employees only work on one project at a time. All employees are **given** a different grade such as, partner, associate, etc. The employees have to account for the time spent and **fill out** time sheets.'

Figure 9.9 The five entities

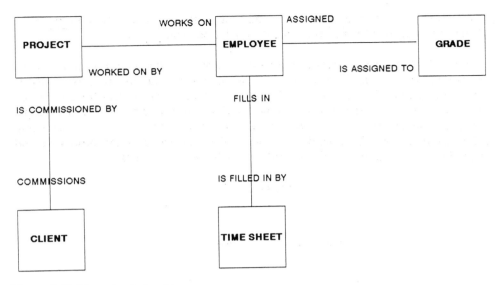

Figure 9.10 Named relationship

The cardinality () and optionality [] are then deduced and inferred from the text and drawn on the diagram, see Figure 9.11.

'(A client) **commissions** a project or (projects) and all projects [must] have a client. The project has (employees) **working on** the project and for management reasons employees only work on (one) project at a time. [All] employees

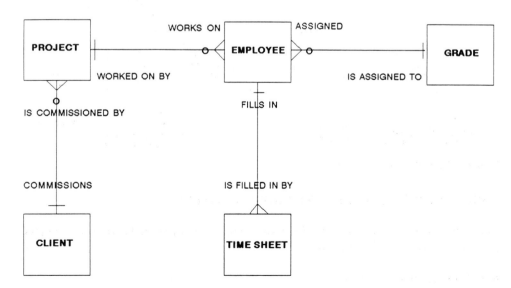

Figure 9.11 Cardinality and optionality

are **given** a different grade such as, partner, associate, etc. The employees [have] to account for the time spent and **fill out** (time sheets).'

Figure 9.12 shows examples of the data, occurrences, that could occur in a real-life situation. The actual occurrence is in large type and the type of entity is contained within the brackets (). The relationship name is positioned on the arrow to show the direction in which the relationship should be read. For example the client called Store PLC commissions two projects to build a new store and refurbish an old one. The build new store project is worked on by John, Jane and Jake. Jane is a quantity surveyor and fills in three time sheets.

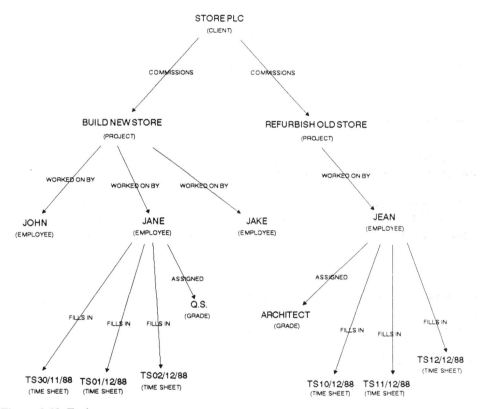

Figure 9.12 Entity occurrences

CASE STUDY EXAMPLE: INFORMATION ANALYSIS

The RICS study developed the following general data models for each subject area for quantity surveying use:

- Organisation structure
- Employee functional role

- Project management
- Project team roles
- Procurement method
- Post contract financial control

These apply to three different types of quantity surveying practice namely Private Practice, Multi-disciplinary Partnership and Local Authority. The latter area has been generalised to cover any public or statutory undertaking. The model developed is called a generic model because it contains the data model for all types of quantity surveying practice. Any integrated databases built using this model as a base would be able to exchange information with each other in a consistent and reliable manner. The data model provides a standard for the data and information required by quantity surveying practices.

Each data model is presented in the same way. An entity–relationship model is drawn supported by a description of the entities in that model. Each diagram corresponds to a subject area and in practice all the entities and relationships would be drawn on one diagram. This has not been done as it is easier to understand the models if they are drawn separately. To obtain a deeper understanding of the model actual occurrences of the entities and relationships are shown for each of the data models. Because of space limitations not all entity type occurrences are shown and only the relationships in the direction of the arrow are labelled.

Organisation Structure

Subject area
Describes the organisational structure of the enterprise, see Figures 9.13(a) and (b).

Entities modelled
ORGANISATION relates to information concerning the organisation as a legal entity.

FUNCTIONAL UNIT is a part of the organisation which performs some identifiable role. It could be the head office, a branch office, a department such as project planning, secretarial services. These functional units could either be centralised or de-centralised.

PROJECT on which the functional unit is spending resources and would include both internal, purely speculative, and external projects for a client.

BUSINESS PLAN is a description of the long to medium term plans of the organisation.

FINANCIAL PLAN is the financial implications of the business plan.

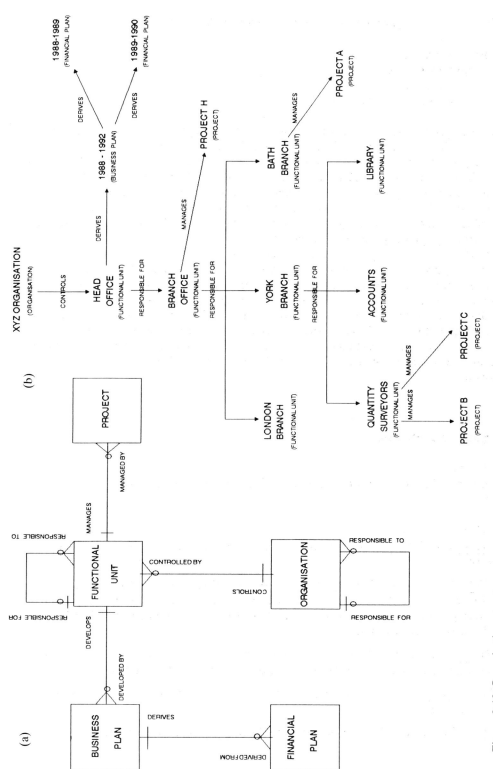

Figure 9.13 Organisation structure

(a)

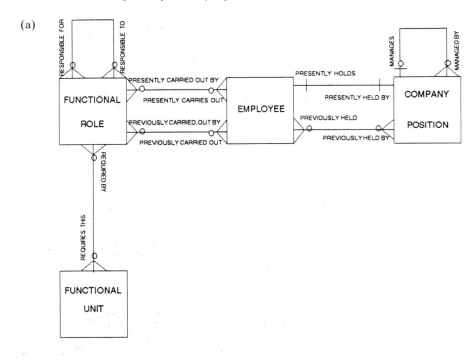

(b)

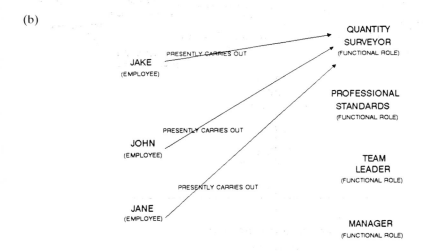

Figure 9.14 Employee functional role

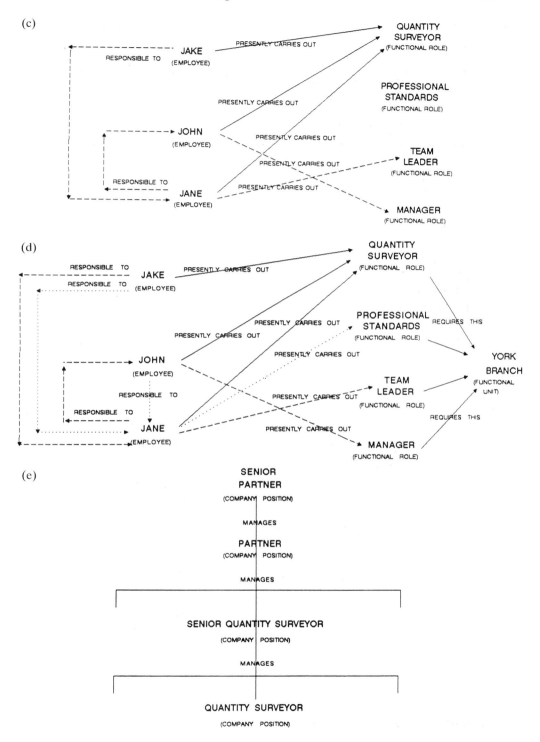

(c)

JAKE (EMPLOYEE)
PRESENTLY CARRIES OUT
RESPONSIBLE TO

QUANTITY SURVEYOR (FUNCTIONAL ROLE)

PROFESSIONAL STANDARDS (FUNCTIONAL ROLE)

JOHN (EMPLOYEE)
PRESENTLY CARRIES OUT
PRESENTLY CARRIES OUT
PRESENTLY CARRIES OUT
PRESENTLY CARRIES OUT

TEAM LEADER (FUNCTIONAL ROLE)

RESPONSIBLE TO

JANE (EMPLOYEE)

MANAGER (FUNCTIONAL ROLE)

(d)

RESPONSIBLE TO
RESPONSIBLE TO
JAKE (EMPLOYEE)
PRESENTLY CARRIES OUT

QUANTITY SURVEYOR (FUNCTIONAL ROLE)

PROFESSIONAL STANDARDS (FUNCTIONAL ROLE)
REQUIRES THIS

PRESENTLY CARRIES OUT
PRESENTLY CARRIES OUT
PRESENTLY CARRIES OUT

JOHN (EMPLOYEE)

YORK BRANCH (FUNCTIONAL UNIT)

TEAM LEADER (FUNCTIONAL ROLE)

RESPONSIBLE TO
PRESENTLY CARRIES OUT
RESPONSIBLE TO
JANE (EMPLOYEE)

PRESENTLY CARRIES OUT

MANAGER (FUNCTIONAL ROLE)

REQUIRES THIS

(e)

SENIOR PARTNER
(COMPANY POSITION)

MANAGES

PARTNER
(COMPANY POSITION)

MANAGES

SENIOR QUANTITY SURVEYOR
(COMPANY POSITION)

MANAGES

QUANTITY SURVEYOR
(COMPANY POSITION)

Comments
The model allows the organisation to be structured as either decentralised/centralised and hierarchical/network. If new firms are acquired or the organisation is taken over the model is flexible enough to continue to be used.

Employee functional role

Subject area
Describes the position of the employee within the organisation, see Figures 9.14(a)–(e).

Entities modelled
FUNCTIONAL UNIT, described previously.

EMPLOYEE is a person who works for the organisation.

COMPANY POSITION is the grade or title the employee holds within the organisation.

FUNCTIONAL ROLE, a responsibility or task that the employee carries has within the organisation.

Comments
Figure 9.14(d) demonstrates the typical hierarchical line management structure. In contrast Figures 9.14(b) and 9.14(c) model a network structure of line management and professional responsibilities. For example Figure 9.14(c) shows that JANE is RESPONSIBLE TO JOHN, who is her MANAGER, but JOHN IS RESPONSIBLE TO JANE for PROFESSIONAL STANDARDS.

One of the major problems that beset large quantity surveying firms is a detailed knowledge of the skills of their employees. The FUNCTIONAL ROLE describes extra skills, not obvious from the description given by the COMPANY POSITION, and could be used to compile registers of skills.

Project Organisation

Subject area
Describes how the project is broken down into stages which worked on by teams form a particular functional unit, see Figures 9.15(a) and (b).

Entities modelled
PROJECT is a project on which the organisation is working. As soon as a project is started it must be given a unique identifier and some basic details. This will allow an analysis to be carried out to determine the status and success of projects.

PROJECT STAGE, each project will be divided into whatever stages are thought appropriate for managing the project.

(a)

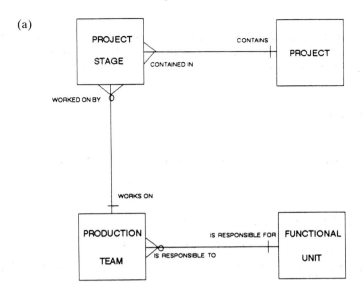

(b)

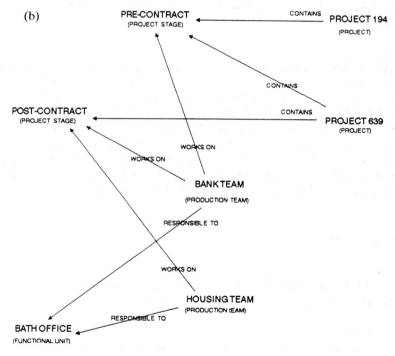

Figure 9.15 Project organisation

PRODUCTION TEAM is the team which is working on the project. It will include members of the organisation, outside consultants, contractors and the client.

FUNCTIONAL UNIT, described previously.

Project Team Roles

Subject area
Describes the roles that have to be carried out by the production team in order to complete the project, see Figures 9.16(a) and (b).

Entities modelled
CLIENT, the person or company that wants the work completed.

TEAM ROLE, a role (task, job) that is required if the work is to be completed.

EMPLOYEE, described previously.

EXTERNAL PROFESSIONAL MEMBER, a person or company employed by the organisation to perform a particular team role or provide a particular expertise not available within the organisation. These are not contract staff who would be considered an employee, albeit with special status, of the organisation.

CONTRACTOR, a person or company who has contracted to execute construction work.

Comments
As mentioned previously large organisations may be unsure of precisely what skills they possess. The TEAM ROLE, essentially a skill description, could be used to form a register not only of employee skills, but of skills available from outside sources.
   In practice each employee would have relationships with entities such as HOLIDAY, TIME CARD, JOB ASSIGNMENT, etc. which could be used to provide detailed information as to their availability for extra work.

Procurement method

Subject area
Describes the different procurement paths that may be adopted (see Figure 9.17).

Entities modelled
FORM OF CONTRACT to be used on the project.

PROJECT STAGE, described previously.

DESIGN, the details of the design.

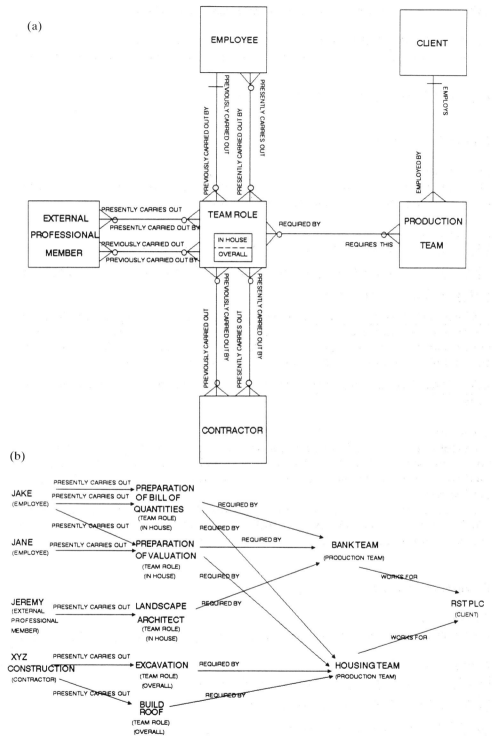

Figure 9.16 project team roles

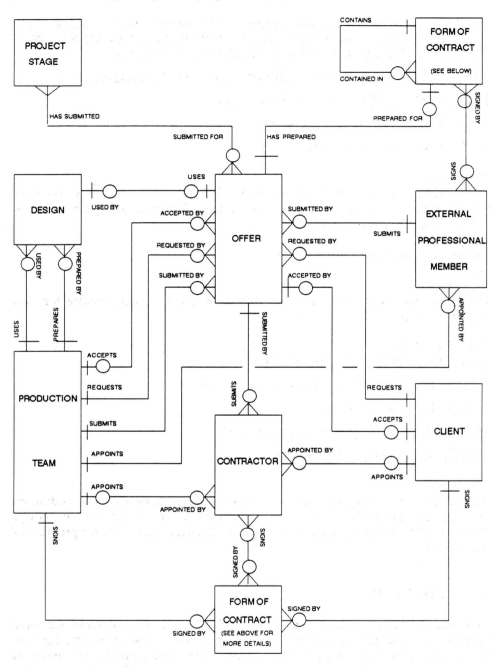

Figure 9.17 Procurement method

TENDER to complete the work. From a CONTRACTOR is called a CON-TRACTOR TENDER and from PRODUCTION TEAM and EXTERNAL PRO-FESSIONAL MEMBER is called a CONSULTANT TENDER.

EXTERNAL PROFESSIONAL MEMBER, described previously.

PRODUCTION TEAM, described previously.

CONTRACTOR, described previously

CLIENT described previously.

Comments
In practice procurement methods do not fall into such neat categories. However the occurrences demonstrate that a wide variety of procurement paths could be accommodated by the model.

Construction management

Subject area
Evaluation for work done and completion of certificates for contractor, see Figures 9.18(a) and (b).

Entities modelled
FORM OF CONTRACT, described previously.

FINAL ACCOUNT, summary of all work done.

VALUATION for certificate.

PRODUCTION TEAM, described previously.

CERTIFICATE authorising payment.

Comments
This is only a very simple model of construction management, concentrating on post-contract financial control. A more detailed model would contain enough detail to enable the project manager to keep a detailed account of the progress of the project.

MODEL OVERVIEW

The model overview is shown in Figure 9.19. The six major models that have been constructed (see Figures 9.13−9.18) are shown as round cornered boxes. The four entities that appear in one or more of the major models are shown as simple rectangles. The double headed arrows show in which major model the entities appear. For example the FUNCTIONAL UNIT entity is contained in the ORGANISATION STRUCTURE, EMPLOYEE FUNCTIONAL ROLE and PROJECT ORGANISATION models. There are a total of 26 entities in the total

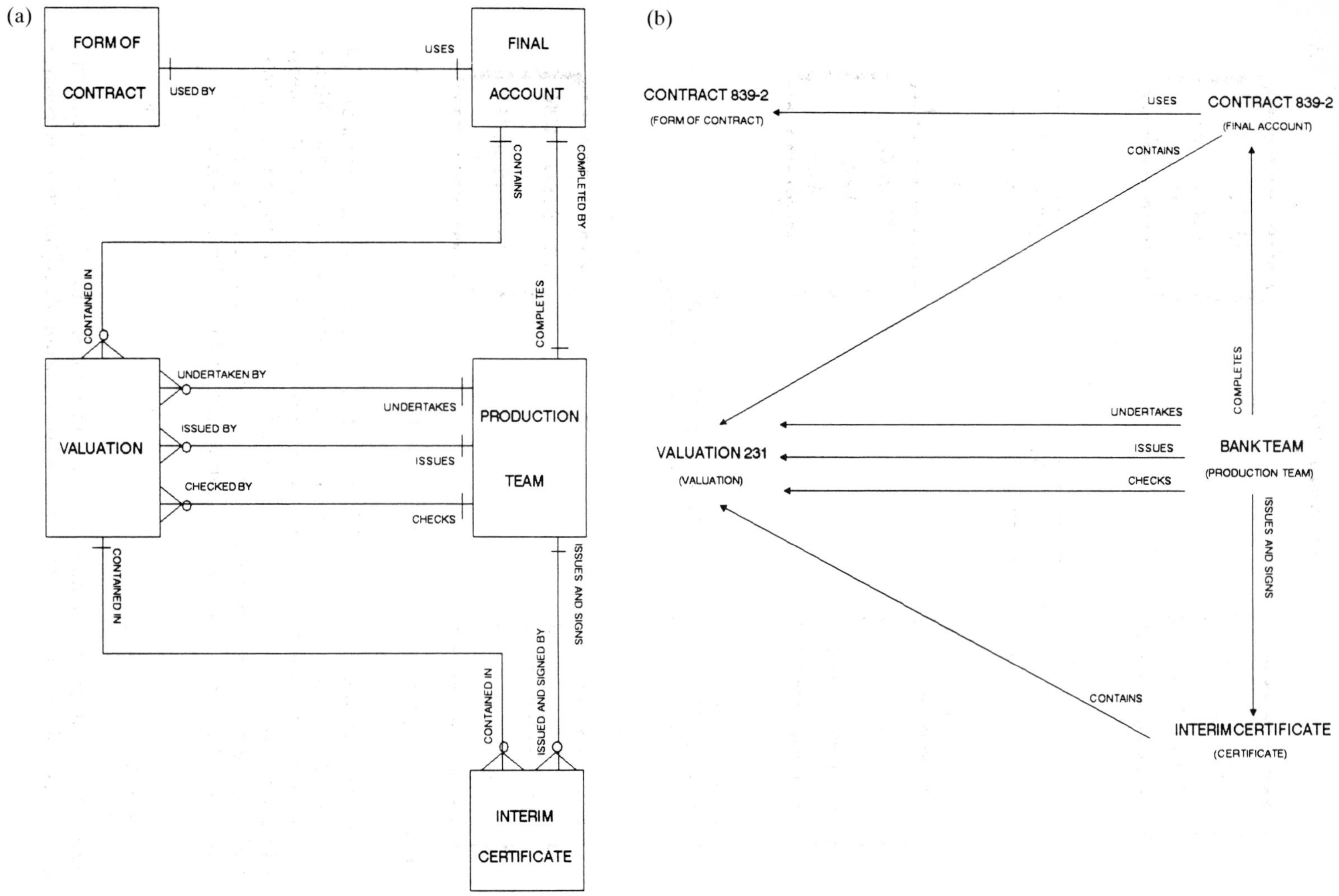

Figure 9.18 Construction management

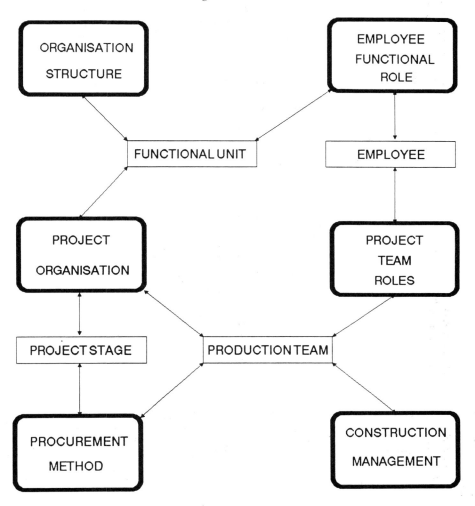

Figure 9.19 Integrated entity−relationship model

model and only four entities (16%) are part of more than one major model. Of the four entities one is in four major models, one in three and two in two. All the major models are not tightly bound to each other yet all form part of a highly integrated database system. This allows the database to be developed over a long period of time in easily manageable stages. The models, subject areas, can be linked together by relating the common entities between the diagrams.

Data analysis

The entity−relationship models developed in Figures 9.13−9.18 are independent of any particular database system. The following example shows how the entity−

relationship model, of Figure 9.15, can be transformed into the relational data model.

The entities of Figure 9.15(a) will have associated with them a number of attributes. With respect to the entity PROJECT there will be a project number, date to be started, date to be finished, etc. For convenience and brevity only attributes and groups of attributes which identify the data will be shown. The entity PROJECT is identified by a project number. In the example in Figure 9.15(b) there are two projects numbered 194 and 639. There would also be further details about the project, such as when it was started, when it should end, etc. These have been contained, for convenience and brevity, in a general description called project details (whose values are represented by '—'). Table 9.1 shows the entity PROJECT and all its associated data.

Table 9.1 Entity PROJECT

Project number	Project details
194	—
639	—

The entity PROJECT STAGE can be developed in a similar manner, see Table 9.2.

Table 9.2 Entity PROJECT STAGE

Project stage	Project stage details
PRE-CONTRACT	—
POST-CONTRACT	—

The two entities PROJECT and PROJECT STAGE have been defined. There are however relationships between them which must be represented. To represent these project is introduced into the entity PROJECT STAGE, see Table 9.3.

Table 9.3 ENTITY PROJECT STAGE

Project stage	Project	Project stage details
PRE-CONTRACT	194	—
POST-CONTRACT	639	—
PRE-CONTRACT	639	—

The entity PRODUCTION TEAM is shown in Table 9.4.

Table 9.4 Entity PRODUCTION TEAM

Production team	Production team details
BANK	—
HOUSING	—

There are relationships between PRODUCTION TEAM and PROJECT STAGE. These are represented by introducing production team into the entity PROJECT STAGE, see Table 9.5.

Table 9.5 Entity PROJECT STAGE

Project stage	Project	Production team	Project stage details
PRE-CONTRACT	194	BANK	—
POST-CONTRACT	639	HOUSING	—
PRE-CONTRACT	639	BANK	—

The entity FUNCTIONAL UNIT is shown in Table 9.6.

Table 9.6 Entity FUNCTIONAL UNIT

Functional unit	Functional unit details
BATH OFFICE	—

There are also relationships between FUNCTIONAL UNIT and PRODUCTION TEAM. These are represented by functional unit in the entity PRODUCTION TEAM, see Table 9.7.

Table 9.7 Entity PRODUCTION TEAM

Production team	Functional unit	Production team details
BANK	BATH OFFICE	—
HOUSING	BATH OFFICE	—

The entity–relationship model of Figure 9.15 has been transformed into a relational database model as shown in Tables 9.1, 9.5, 9.6 and 9.7. This is still independent of any particular database software such as dBase III Plus, ORACLE, etc.

IMPLEMENTATION USING FOURTH GENERATION LANGUAGES

A fourth generation language is another name for a database management system. The conceptual schema is described by a simple form filling exercise (as in dBase III Plus) or by a data definition language (ORACLE) and then determines how to store the data (internal schema). The information can then be accessed by simple form filling (as in dBase III Plus) or a data manipulation language such as SQL (ORACLE).

dBase III Plus

The Tables 9.1, 9.5, 9.6 and 9.7 could be implemented as dBase III Plus database files as shown in Tables 9.8, 9.9, 9.10 and 9.11 respectively. The database files are described as follows:

Field Name	The name of the attribute within the entity.
Type	The type of data the attribute will hold. A 'numeric' attribute will

	hold numbers (digits) and 'character' attributes alphabetic characters and numbers.		
Width	The maximum number of digits and/or characters held in the attribute.		
Dec	The number of decimal places held if the type is numeric.		

Table 9.8 PROJECT database file

	Field name	Type	Width	Dec
1	PROJECT	Numeric	5	
2	PROJECTDETAILS	Character	80	

The type and length of PROJECT in the database files in Tables 9.8 and 9.9 are the same.

Table 9.9 PROJECT STAGE database file

	Field name	Type	Width	Dec
1	PROJECTSTAGE	Character	20	
2	PROJECT	Numeric	5	
3	PRODUCTIONTEAM	Character	10	
4	PROJECTSTAGEDETAILS	Character	80	

The type and length of PRODUCTION TEAM in the database files of Tables 9.9 and 9.11 are the same.

Table 9.10 FUNCTIONAL UNIT database file

	Field name	Type	Width	Dec
1	FUNCTIONALUNIT	Character	20	
2	FUNCTIONALUNITDETAILS	Character	80	

Table 9.11 PRODUCTION team database file

	Field name	Type	Width	Dec
1	PRODUCTIONTEAM	Character	10	
2	FUNCTIONALUNIT	Character	20	
3	PRODUCTIONTEAMDETAILS	Character	80	

The type and length of FUNCTIONAL UNIT in the database files in Tables 9.10 and 9.11 are the same.

The database files have now been developed. It is now a simple matter to define the procedures which access the data.

ORACLE

The Tables 9.1, 9.5, 9.6 and 9.7 can also be implemented as ORACLE database tables as shown in Tables 9.12, 9.13, 9.14 and 9.15 respectively. In this example the further details have been expanded. For example in the PROJECT the PROJECT DETAILS have been expanded to include PROJECT DESCRIPTION, PROJECT START DATE, PROJECT END DATE and PROJECT CONTACT. The attributes are described in a similar manner to dBase III Plus where 'char' corresponds to character and 'number' to numeric. The lengths are contained within the brackets. The text between the '/*' and the '*/' are treated as comments to explain and annotate the definition of the database tables.

Table 9.12 PROTECT TABLE

```
/*                                                                    */
/*    ------------------------------------PROJECT  TABLE-------------------------------------    */
/*                                                                    */
create table project (
    project              number(5)      not null,
    project_description  char(80)       not null,
    protect_start_date   date,
    project_end_date     date,
    project_contact      char(30))
```

The 'project' is 'not null' because it uniquely identifies the project and must always have a value. The 'project_description' is 'not null' because a value must always be entered so that users understand what the project is all about.

Table 9.13. PROJECT STAGE TABLE

```
/*                                                                    */
/*    ---------------------------------PROJECT  STAGE  TABLE-----------------------------    */
/*                                                                    */
create table project_stage (
    project_stage        char(20)       not null,
    project              number(5)      not null,
    production_team      char(10)       not null,
    stage_description    char(50)       not null,
    stage_start_date     date,
    stage_end_date       date)
```

The 'project_stage', 'project' and 'production_team' are 'not null' because they uniquely identify the project stage and must always have a value. The 'stage_description' is 'not null' because a value must always be entered so that users understand what the project stage is all about.

Table 9.14 FUNCTIONAL UNIT TABLE

```
/*                                                                    */
/*    -----------------------------FUNCTIONAL  UNIT  TABLE-----------------------------    */
/*                                                                    */
create table functional_unit (
    functional_unit      char(20)       not null,
    unit_location        char(30),
    unit_function        char(20))
```

The 'functional_unit' is 'not null' because it uniquely identifies the functional unit and must always have a value.

Table 9.15 PRODUCTION TEAM TABLE

```
/*                                                                        */
/*        ----------------------------PRODUCTION  TEAM  TABLE----------------------------   */
/*                                                                        */
create table production__team (
       production__team              char(20)        not null,
       functional__unit              char(30)        not null,
       production__team__details     char(80))
```

The 'production__team' and 'functional__unit' are 'not null' because they uniquely identify the production team and must always have a value.

The tables have now been developed. It is now a simple matter to define the SQL statements to access the data.

A simple example of a screen is shown in Figure 9.20.

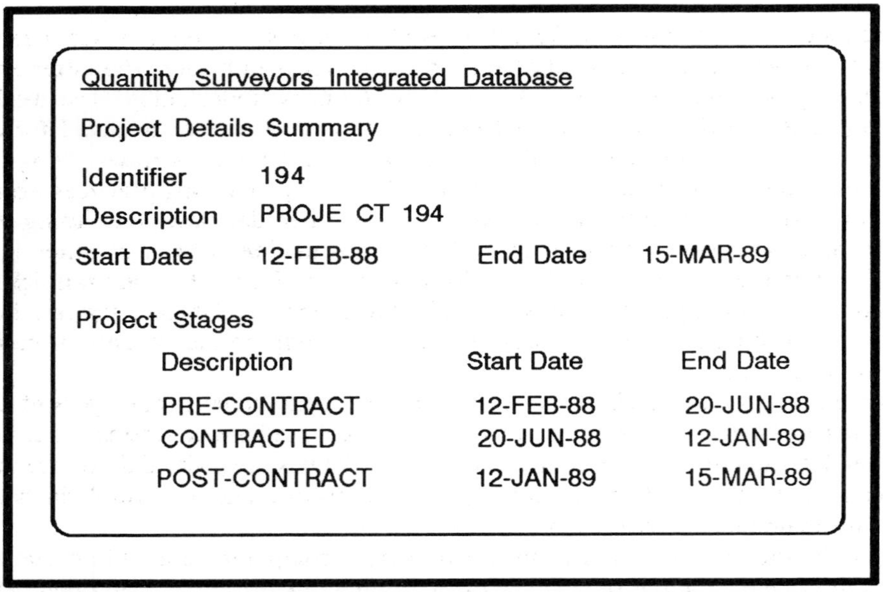

Quantity Surveyors Integrated Database		
Project Details Summary		
Identifier 194		
Description PROJE CT 194		
Start Date 12-FEB-88	End Date 15-MAR-89	
Project Stages		
Description	Start Date	End Date
PRE-CONTRACT	12-FEB-88	20-JUN-88
CONTRACTED	20-JUN-88	12-JAN-89
POST-CONTRACT	12-JAN-89	15-MAR-89

Figure 9.20 Typical output screen

CONCLUSIONS

Knowledge is a major source of competitive advantage as the application of information technology removes national and international boundaries making information more freely available. It is important that quantity surveyors develop information structures (integrated databases), support and decision systems to enable them to take advantage of these opportunities. This would enable a value added service to be provided to clients and the management focus which the 'holder' of such key data would attract and clearly strengthen the quantity surveyor's position in the design team. A method, information engineering, for the development of an integrated database for quantity surveyors is described. This approach emphasises the importance of determining what information is required by the quantity surveyor. Once this information, which is presented in graphical form, has been determined a variety of strategies are suggested which could be used to build the database.

The following general data models have been developed:

- Organisation structure
- Employee functional role
- Project management
- Project team roles
- Procurement method
- Post-contract financial control

These models apply to three different types of quantity surveying practice namely private practice, multi-disciplinary partnership and local authority. The latter area has been generalised to cover any public or statutory undertaking. The model developed is called a generic model because it contains the data model for all types of quantity surveying practice. This model, it must be stressed, is not a complete model. A complete model for typical quantity surveying practices would contain more than 500 entities. However, any integrated databases built using this model as a base would be able to exchange information with each other in a consistent and reliable manner. The major models, see Figure 9.19, are not tightly bound to each other yet all form part of a highly integrated database system. This would allow the database to be developed over a long period of time in easily manageable stages.

There is at present a lack of awareness amongst many quantity surveyors of the advantages of using integrated databases. At a recent conference session, attended by over 100 quantity surveyors, it was clear that the vast majority did not use any methods or tools to help them develop their software. This will result in large software maintenance costs in the future.

Quantity surveyors considering the need to apply computers to their information needs are urged to adopt the principles of database development described. The decision on the next step will clearly be one that can only be taken by the organisation itself and will be influenced by the size and complexity of the database required, the resources of the organisation and the knowledge available. The solution could lie anyway on the path from simple 'off the shelf' commercial packages to the development of a bespoke system, possibly using consultancy assistance.

The RICS study can be seen as a first step towards developing an integrated database solution for quantity surveyors (or possibly the industry as a whole). If further work proceeds then clearly the results of this would also be available to organisations. It would have the advantage of providing the quantity surveyor with greater business efficiency and the possibility of providing a more effective service to clients. It would also lay down a coherent framework which would aid the more effective use and communication of data within the construction industry.

ACKNOWLEDGEMENTS

The authors wish to acknowledge that this part of the manual draws upon a study undertaken by Salford University for the Royal Institution of Chartered Surveyors QS Research and Development Committee. They would like to thank the organisations who participated in the case study and workshop sessions, Jane Winn of Oracle Corporation UK for developing the ORACLE tables and the output screen and James Martin Associates for introducing them to the information engineering method.

BIBLIOGRAPHY

Brandon, P. and Kirkham, J. A. (1989) *An Integrated Database for Quantity Surveyors*. London: Surveyors Publications, RICS.

Deen, S. D. (1977) *Fundamentals of Data Base Systems*. London: Macmillan Computer Science Series.

Martin, J. (1987) *Information Engineering*, Volume I, II and III. Carnforth: Savant Research Studies.

Oracle (1988) *SQL The Quiet Revolution*. Bracknell: Oracle Systems Software B.V.

Chapter 10

Procurement Systems for Building

JOHN BENNETT, *Professor of Quantity Surveying,*
University of Reading and TONY GRICE, *Bucknall Austin plc*

INTRODUCTION

The choice of an appropriate procurement system is crucial to the success of
building projects. As the Building Economic Development Council's *Faster Building
for Commerce* (1988) makes clear, the penalties of a poorly conceived and badly
run project are:

(1) longer periods with capital tied up, incurring extra bank borrowing charges or
    loss of return;
(2) loss of business;
(3) greater uncertainty in managing business assets when opening dates or letting
    dates cannot be fixed;
(4) opportunities for the competition to get ahead.

The procurement system establishes the roles and relationships which make up
the project organisation. It establishes the overall management structure and
systems. It helps shape the overall values and style of the project. The choice of
procurement system is a crucial strategic decision equalled only by the establishment
of the client's objectives and deciding the nature of the end product. All three key
strategic elements, objectives, end product and procurement system, have a
decisive influence on the level of success achieved by building projects.

Making an appropriate choice requires that the client's objectives and the
essential nature of the required end product are analysed very early in the life of
the project. On the basis of this analysis a procurement route is selected. Then
throughout the project, care must be taken to ensure that the roles and relationships
established, the project management approach, the pattern of meetings and
information systems, the form of contracts, and the overall style and values of the
project organisation are all consistent with the selected procurement route.

The choice of an appropriate procurement system in the UK in the early 1990s
is conditioned by some important features of the national building industry. First,
the building industry is unusually fragmented. Any one building project requires a
large number of specialists employed by separate firms to be welded together into
a single, effective temporary organisation. The specialists include consultants
(architects, engineers and surveyors) and contractors (general, management and
specialist).

Secondly, specialists are selected on the basis of competition and are usually employed under contracts which tend to emphasize the conflict of interests between client and specialist rather than seek to build a co-operative, co-ordinated project organisation.

Thirdly, the role of specialist contractors is changing. Virtually all direct construction work is provided by separate specialist contractors. As their specialisms become more esoteric (mainly in industrial and engineering based trades) so they become responsible for detail design and many aspects of the site management of their own work. In recognition of these trends, most procurement systems now include some form of contractual link between the client and at least the key specialist designers.

Fourthly, British architects normally have a contractual right and responsibility to be involved in all aspects of detail design. This creates difficulties in managing design and in establishing design liabilities when the knowledge needed to make most design decisions resides outside of the architect's own firm.

Fifthly, and largely in response to difficulties caused by the first four features of the UK building industry, there is now wide acceptance of the importance of management within building projects. It must also be said that this is a comparatively recent phenomenon and consequently there is fierce competition for the various management responsibilities arising within modern building projects. Almost all the various specialists in the industry, at one time or another, lay claim to a management role, if only in respect of certain categories of project.

However the noteworthy feature is that management has emerged alongside design and construction as a basic and fundamental responsibility within building projects.

Finally, expert and experienced clients now play a more active role in building projects than in earlier times. They are not content merely to set tough targets for time, quality and cost, they now become involved in all aspects of their projects. Clients challenge all aspects of the building industry's performance in a search for better value, faster construction and bigger profits. In this way the client's role has become very important in building procurement.

The key features of fragmentation, conflict, changing roles, uncertainty over design liability and management responsibility combined with proactive and impatient clients make the choice of an appropriate procurement system for building projects both difficult and very important.

PROCUREMENT OPTIONS

There are many different procurement systems and several ways of classifying them. The list used by the Building EDC in *Thinking about Building* provides a sensible list of the options:

(1) Traditional — the client appoints consultants for design and cost control and later selects a main contractor to carry out the work. Some contractors work for an agreed fee plus the actual costs of the direct construction work rather than for a lump sum which includes an undisclosed profit. Two variants are:

(a) Sequential — contractors bid on completed design and cost documents.
(b) Accelerated — a contractor is appointed early on the basis of partial information, by negotiation or in competition, possibly on a two-stage basis.

(2) Design and build — the client buys the completed building from a contractor who is responsible for design and construction. The approach is referred to as a turnkey contract when it includes the complete equipping and/or staffing and commissioning of a building. Three variants are:

(a) Direct — a designer-contractor is appointed after some appraisal but without competition.
(b) Competitive — documents are prepared by consultants to enable several contractors to offer designs and prices in competition.
(c) Develop and construct — consultants are appointed to design the building to a partial stage, then contractors complete and guarantee the design in competition, either using the client's consultants or their own designers.

(3) Management — the client appoints design and cost consultants and a contractor or consultant to manage construction for a fee. Specialist contractors are appointed to undertake the construction work by negotiation or in competition. Two variants are:

(a) Management contracting — a management contractor takes some contractual risks in delivering the project to an agreed price and on time and employs the specialist contractors as sub-contractors. Although the price may be guaranteed, this is unusual, and clients retain some time and price risks.
(b) Construction management — a professional firm is paid a fee to provide the management service and the specialist contractors enter into direct contracts with the client, who retains the time and price risks.

(4) Design and manage — the client appoints a single firm to design and deliver the project, but specialist contractors are appointed to undertake the construction work by negotiation or in competition. Two variants are:

(a) Contractor — the project design and manage firm takes a contractual risk in delivering the project to an agreed price (which may be guaranteed) and on time and employs design consultants and specialist contractors as sub-contractors.
(b) Consultant — the project designer and manager is employed as the client's agent and the specialist contractors enter into direct contracts with the client, who retains the time and price risks.

There are important differences between the procurement systems available. They are best considered from the view-point of the client because this serves to highlight the key choices which need to be made in selecting an appropriate option.

Quality, cost and time

The most important issues for clients concern quality, cost and time. All the options can provide satisfactory performance in all these matters but they provide different emphases and different levels of risk and control for clients. Figure 10.1 provides a broad indication of the strengths and weaknesses of the procurement options. This shows, for example, that clients who are content with straightforward reliable quality and want economy and speed should consider the design and build approach. For clients who want a more individual design and higher performance and are prepared to pay more and wait longer, the traditional general contractor approaches are worth considering. On the other hand the management based approaches are flexible. They can provide for innovative, original design quality, fast completion and can achieve economy. However there are trade-offs between innovation, originality, quality, fast construction, an early start on site and the consequential costs. So clients need to consider, with their professional advisors, the weight they place on specific values of quality, cost and time.

Figure 10.1 provides space for clients to record the weights which reflect their priorities. The use of the form to select an appropriate procurement system is explained later in the chapter. The utility factors allocated to each procurement system are the authors' judgement of sensible values. They take account of the Building EDC's *Thinking about Building* (1985) and Skitmore and Marsden's (1988) extension of the scheme which it provides. Before demonstrating the use of the form it is necessary to consider some important issues.

Quality

In considering quality at the early stages of projects, clients with their advisors need to select the overall level of quality. That is the choice between rare, unusual, high performance materials, great precision and sophistication and hand crafted excellence on the one hand and straightforward, basic competence on the other; i.e. quality is defined as being fit for the purpose as perceived by the client to suit his particular needs. In simple terms clients need to make a choice between a Rolls Royce and a Mini.

It is taken for granted that, whatever level of quality is selected, the project organisation will provide a proper system of quality assurance to ensure that the required performance is achieved.

Single-point responsibility

Once the key issues of quality, cost and time are resolved the biggest choice for clients is whether they want to be closely involved in the strategic management of

Procurement systems

Client's priority: Essential 5 Desirable 4 3 2 Do without 1	Traditional				Design & Build						Management				Design & Manage			
	Sequential		Accelerated		Direct		Competitive		Develop & Construct		Management contracting		Construction Management		Contractor		Consultant	
	Utility	Score	Utility	Score	Utility	Score	Utility	Score	Utility	Score	Utility	Score	Utility	Score	Utility	Score	Utility	Score
TIME Is early completion required?	10		50		100		90		60		100		100		90		80	
COST Is a firm price needed before any commitment to construction is formed?	90		40		100		100		90		20		10		30		20	
FLEXIBILITY Are variations necessary after work has begun on site?	100		90		30		30		40		80		90		60		70	
COMPLEXITY Is the building highly specialised, technologically advanced or highly serviced?	40		20		20		10		40		100		100		70		80	
QUALITY Is high quality important?	100		60		40		40		70		90		100		50		60	
CERTAINTY Is completion on time important?	50		30		100		90		70		90		90		100		90	
Is completion within budget important?	30		30		100		100		50		70		60		90		90	
DIVISION OF RESPONSIBILITY Is single - point responsibility wanted?	30		30		100		100		70		30		10		90		90	
Is direct professional responsibility wanted?	100		100		10		10		50		70		100		30		30	
RISK Is transfer of responsibility for the consequence of slippages important?	30		30		80		100		70		30		10		100		80	
Results																		

Figure 10.1 Strengths and weaknesses of procurement systems

their project or simply to appoint a single firm to take this responsibility. As Figure 10.2 shows, the extremes are represented by construction management which requires a considerable amount of detailed client involvement, and at the other end of the spectrum, design and build provides single-point responsibility. This allows clients to concentrate on defining their requirements and then merely monitor the contractor's decisions. All the options between the two extremes, which require the direct employment of more than one firm expose clients to problems arising from the need to co-ordinate divided responsibilities. These can include inconsistencies in design information, late design information, misunderstandings between designers and constructors, uncertainty over construction method and the knock-on effects of late construction work. On the other hand direct contracts with consultants, contractors and specialists provides clients with considerable control and influence over their projects. They can be closely involved in selecting firms which understand their own values and style. They can ensure that all the essential specialist knowledge and skills are brought into their projects at the most effective stage. Thus for example construction experience and knowledge can be made available to improve the economy and efficiency of design decisions by the early involvement of contractors.

Construction commitment

The next important choice is the stage in the complete design and construction process at which the client is able to enter into a firm lump sum contract for the direct construction work. Figure 10.3 shows that this varies from the very earliest stages when using the design and build approach. Various forms of the other options allow firm construction contracts to be formed throughout all of the design stages. Then finally the management based approaches do not require the client to enter into a firm commitment to construction until the individual specialist contractors are appointed.

An issue which is often included in lists of criteria to be considered when selecting a procurement system, is whether the client needs to select contractors by competition. This is an important issue but all the procurement systems allow clients to negotiate with one carefully selected contractor or use competition.

Flexibility

The next major issue is the facility with which the client can order changes during the design and construction process. All the options give clients the authority to order changes but the more important issue is how easily and painlessly, from the client's viewpoint, can the consequences on costs and times be established. Figure 10.4 shows that changes are most difficult for clients using design and build. On the other hand the management based approaches provide the greatest flexibility for clients. In practice many of the most ingenious approaches, including, for example, bills of approximate quantities, two-stage tenders and target cost contracts,

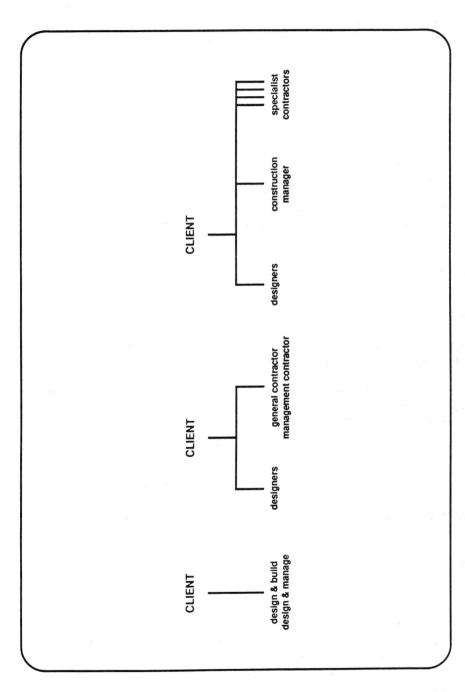

Figure 10.2 Single-point or fragmented responsibility

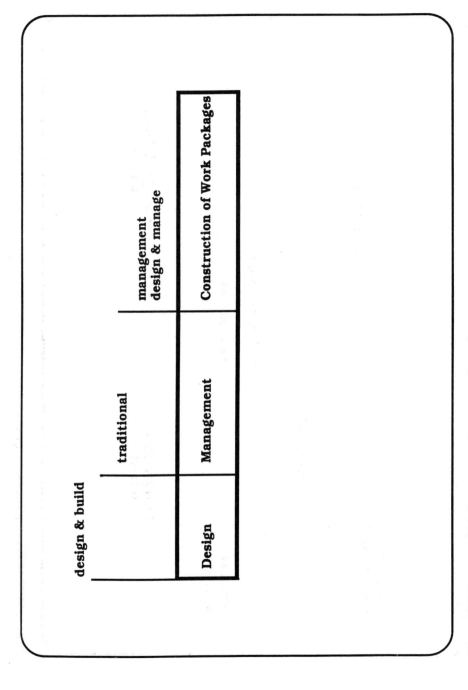

Figure 10.3 Stage at which firm construction contract formed

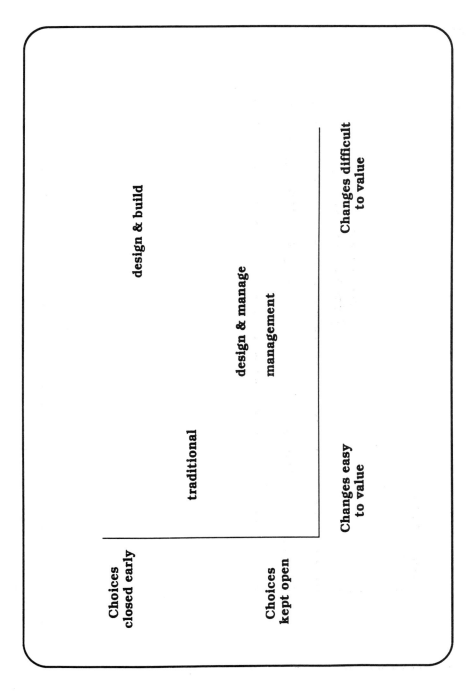

Figure 10.4 Flexibility of procurement systems

have been devised to provide flexibility for clients. Also probably the main advantage of bills of quantities, whether used on traditional contracts or management type contracts, is that they provide an open, contractual basis for valuing changes.

Certainty and risk

Certainty of price and time are closely linked to the degree of flexibility available for clients. Design and build can provide complete contractual certainty on completion for clients from the very earliest stages of their projects. It is obviously true that this certainty is undermined if the client orders changes. However provided there are not many changes and decisions are made in time to avoid interrupting the design, management, manufacturing and construction processes, design and build exposes clients to few risks. At the opposite extreme the management based approaches generally require clients to carry cost and time risks at least until the separate trade construction contracts are let. In this respect the management based approaches leave clients with more of the risks than the alternatives.

SELECTING A PROCUREMENT SYSTEM

The choice of procurement system depends on the client's objectives in terms of function, quality, value, time, cost and certainty. Thus at the second level of analysis, the choice depends on key features of the end product and of the design, management, manufacturing and construction processes required to realise the client's objectives.

Thinking about Building provides a list of issues which must be considered and relates a range of answers to the various procurement systems available. The approach is developed further in Skitmore and Marsden's paper (1988). This developed approach is used as the basis of Figure 10.1 which provides a basis for selecting an appropriate system for any specific building project.

The form should be used by the client with his advisors to record the weight to be given to each of the project objectives. The client's priority should result from a full discussion of the issues involved.

Prior to that meeting the utility factors allocated to each procurement system should be reviewed in the light of the project and the client. It may be that slightly different questions, or additional questions and different utility factors apply to the circumstances of a particular case.

Figure 10.5 shows a completed example of Figure 10.1. It includes a client's priorities and the resultant values given to the procurement systems. The client's priorities in the example imply a high quality, innovative project in which cost and time constraints are relatively unimportant. The selection procedure identifies construction management as the appropriate choice. This is a sensible choice. The procedure indicates that management contracting also provides a reasonable approach. It also suggests that the design and manage systems and the sequential

Procurement systems

Client's priority:
Essential 5
Desirable 4
3
2
Do without 1

Question	Client's priority	Traditional				Design & Build						Management				Design & Manage			
		Sequential		Accelerated		Direct		Competitive		Develop & Construct		Management contracting		Construction Management		Contractor		Consultant	
		Utility	Score	Utility	Score	Utility	Score	Utility	Score	Utility	Score	Utility	Score	Utility	Score	Utility	Score	Utility	Score
TIME Is early completion required?	2	10	20	50	100	100	200	90	180	60	120	100	200	100	200	90	180	80	160
COST Is a firm price needed before any commitment to construction is formed?	2	90	180	40	80	100	200	100	200	90	180	20	40	10	20	30	60	20	40
FLEXIBILITY Are variations necessary after work has begun on site?	5	100	500	90	450	30	150	30	150	40	200	80	400	90	450	60	300	70	350
COMPLEXITY Is the building highly specialised, technologically advanced or highly serviced?	5	40	200	20	100	20	100	10	50	40	200	100	500	100	500	70	350	80	400
QUALITY Is high quality important?	5	100	500	60	300	40	200	40	200	70	350	90	450	100	500	50	250	60	300
CERTAINTY Is completion on time important?	3	50	150	30	90	100	300	90	270	70	210	90	270	90	270	100	300	90	270
Is completion within budget important?	2	30	60	30	60	100	200	100	200	50	100	70	140	60	120	90	180	90	180
DIVISION OF RESPONSIBILITY Is single - point responsibility wanted?	1	30	30	30	30	100	100	100	100	70	70	30	30	10	10	90	90	90	90
Is direct professional responsibility wanted?	3	100	300	100	300	10	30	10	30	50	150	70	210	100	300	30	90	30	90
RISK Is transfer of responsibility for the consequence of slippages important?	3	30	90	30	90	80	240	100	300	70	210	30	90	10	30	100	300	80	240
Results			2030		1600		1720		1680		1790		2330		2400		2100		2120

Figure 10.5 Examples of procurement system selection

version of traditional general contracting deserve some thought. The procedure clearly eliminates design and build and the accelerated traditional systems. Armed with this guidance the client and his advisors should review the decisions which have shaped the results and then make a decision.

Having made the selection it is crucial to the success of the project that the subsequent stages are managed in a manner which supports and is consistent with the initial choice. This requires discipline from the client, consultants and contractors. The following sections describe the key factors required for the success of each of the main types of procurement system from the client's viewpoint.

PROCUREMENT SYSTEMS IN PRACTICE

Traditional

The traditional system, as its name implies, relies on the separate professional disciplines working within established procedures. Thus the use of standard forms of contract, standard methods of measurement and co-ordinated project information are all essential to the smooth functioning of the traditional system. It is essentially a sequential approach. Therefore the client must allow time for all the professions to play their full part in the correct sequence.

The traditional system relies on the use of well understood forms of construction. This is the case because it is unrealistic to ask contractors to give a firm lump sum price for unfamiliar or innovative construction work.

The traditional system provides a basis for efficient construction when the contract is based on well considered and complete project information and the client is determined not to allow the agreed design to be altered. While the traditional forms of contract make extensive provision for variations, the uncertainty generated by excessive change inhibits efficiency and may well leave the client with a building of poor quality, delivered late at a high price.

When speed is important, the accelerated traditional system provides a possible answer. It is most likely to be successful if a project office is created in which the whole project team work together. Otherwise the need for speed will prevent all the professions having the time to make well considered inputs in the right sequence.

Design and build

The two main factors which determine the success of the design and build procurement systems are the client's brief and the quality assurance procedures. It is essential that clients state all their requirements before entering into a design and build contract. The JCT with Contractor's Design form of contract (JCT 81) allows this statement to be as detailed as necessary to reflect those matters which the particular client regards as important. Equally the contract allows for the elements in which the client has no particular requirements, to be defined in performance terms. These performance statements may be very simple, in effect asking merely for a competent answer. Thus the client's brief can be short, leaving much discretion to the contractor and so allowing him to concentrate on producing the most efficient design. Or the brief may be very detailed, leaving only hidden or insignificant elements to the contractor's decision. In either case the attraction of the JCT 81 form is that the contractor is responsible for delivering a complete building for a firm price and to a fixed completion date.

The overall performance depends, in practice, on good quality assurance procedures. These should be linked closely to the brief. At each stage of the complete design, manufacturing and construction process, the client's advisors should check the proposals, the components or the complete building for conformance to the standards defined in the brief. The more the defined answers or required performance can be expressed in terms of objective measurements, the more likely is the success of the project. Much research is needed to provide comprehensive measures for all building elements and performance factors. Meanwhile and until such research is carried out, an effective approach is to require specific features to be equal to those in an identified existing building, preferably one with which the contractor has been involved.

It follows from the two requirements for success, that design and build is unsuitable for complex, innovative projects. It also follows that clients should not change the design because the system provides no equitable basis for valuing variations.

Management

The most distinctive feature of the management approaches to procurement is the central importance of the client's responsibilities. As with all the procurement systems, the client must define the fundamental objectives of his project, select the procurement system and the firms to carry-out the first-line responsibilities, exercise the ultimate authority over the project team and provide the finance. In addition the management approaches require the client to act as chairman of key project meetings. Management contracting and even more so construction management, create teams of equals. Therefore the client must be fully involved in key decisions to ensure that all the participants are allowed to play a full role and also to ensure that his own interests remain paramount. These are demanding responsibilities which deserve very careful thought and preparation by the client before he embarks on the management procurement systems.

An equally important issue is that specialist contractors are given substantial responsibilities. Their design and even more their management role is much larger than in the traditional system. It is therefore necessary to select specialist contractors with previous directly relevant experience or to provide tailored induction courses for all the specialists' workers. Induction needs to deal particularly with the importance of everyone involved taking responsibility for the overall co-ordination of work. This applies both to design and construction. It needs to instil in all workers a 'can-do' attitude when faced with potential problems. Success depends on everyone being actively committed to the project's objectives and refusing to allow obstacles to delay or compromise their performance.

Design and manage

There is little experience of design and manage in the UK. It provides a synthesis of the design and build and management procurement systems. As such it potentially has the strengths and weakness of both. The performance actually achieved will probably depend a great deal on the client defining his objectives clearly and ensuring that they are backed up by effective quality assurance.

Design and manage could emerge as a very significant procurement system. It combines single-point responsibility, which is attractive to many clients, with the organisational flexibility needed to deal with today's varied technologies.

QUANTITY SURVEYORS' ROLE

The modern quantity surveyor plays many roles based on the wide spectrum of knowledge and skills described in the other chapters of this book. These roles may be divided into two broad categories. The first is essentially an advisory and problem-solving role, working as part of a team to define the objectives, the outline design and the procurement system. The second is essentially a formal contractual management role. This most usually comprises the exercising of financial control to ensure that the client's budgetary objectives are achieved.

Both roles are important in all the procurement systems. However their relative importance varies considerably. Figure 10.6 illustrates the different weight given to the problem-solving stages and the management-control stages within the various procurement systems.

It is important to recognise the different style and general behaviour required by the two different stages. The first is a free-ranging, problem-solving process. Heirs and Pehrson provide excellent advice on how organisations should structure these early stages of projects into four distinct steps:

(1) The question. The development of a carefully defined question.
(2) The alternatives. The development of alternative answers by those individual members of the organisation whom management considers most skilled at assembling the needed information and creating or formulating the required alternatives or options.
(3) The consequences. The creation and analysis (by the same or different people) of predictions as to the consequences for the organisation of acting on each alternative.
(4) The judgment/decisions. The use of judgment by the designated decision makers to determine which of the alternatives — or which combination of thoughts among the alternatives presented — should be chosen to be acted on or tested, or what further thinking efforts need to be made in earlier stages of the decision-thinking process before a final decision can be reached.

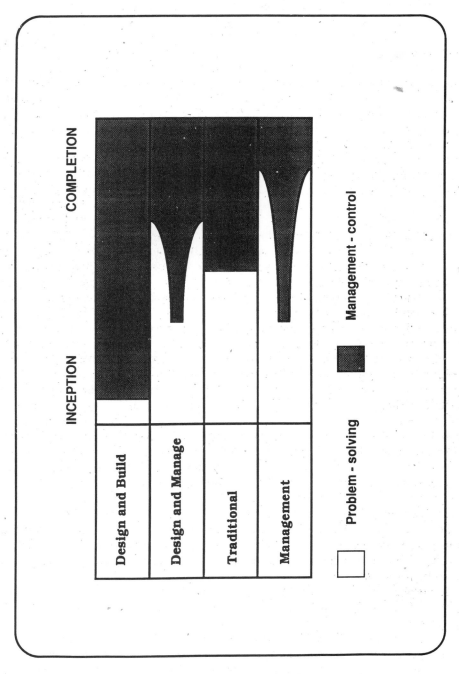

INCEPTION   COMPLETION

Design and Build

Design and Manage

Traditional

Management

Problem - solving

Management - control

Figure 10.6 Stages of the quantity surveyor's role

The second management-control stage is much more formal. It relies on the orderly carrying out of predefined tasks within an agreed schedule and budget. The Engineering Construction EDC's *Guidelines for the Management of Major Projects in the Process Industries*, despite its title, provides good advice applicable to the management-control stages of all types of construction projects. Key points for quantity surveyors include:

(1) Programme. The programme needs to be an honest statement of what it is intended to make happen. The overall programme provides key milestone dates.
(2) Budget. The budget needs to be an honest statement of what it is intended to make happen. It provides individual budgets for each separate contract which make up the complete project.
(3) Design programme and budget. All design decisions should be planned in detail taking account of the needs of procurement, manufacturing, construction and commissioning.
(4) Construction programme and budget. The programme and budget must be realistic and supported by simple and effective communication to all project participants.
(5) Individual programmes and budgets. Every firm involved should have a detailed programme and budget linked to the key dates and budgets of the overall project.
(6) Quality plans. The quality plans of all participants need to be co-ordinated in order to streamline the monitoring and inspection processes.
(7) Contracts. Contracts need to be designed to create co-operative working relationships.
(8) Short-term programmes. Short-terms programmes should deal with the next day, three days or one week, depending on the tempo of the work.
(9) Walk the job. All managers need to walk the design offices, manufacturing plants and construction sites to get a feel for what is really happening, to ask questions and to identify and prevent problems.
(10) Information systems. Information systems should be kept simple and must focus on the key project values — if quality is important, require people to report regularly on quality.

GOOD PRACTICE

Effective procurement depends on clients and the industry itself applying a number of key principles. The Building EDC's studies of industrial and commercial building have identified many of these. The following lists draw on the EDC's work and other research listed at the end of the chapter.

The key principles for clients emphasize the absolutely central importance of this role:

(1) Be committed. You initiate the project, you set the style and tone, and you are an essential member of the team.

(2) Define your role and equip yourself to carry it out. How much do you want to be involved in the process of designing and managing the project? Have you got the resources and expertise to provide the input you want to make? Lean resources in calibre and numbers of staff, especially on site, and economies in overhead costs result in low productivity, failure to meet time targets and waste.

(3) Determine priorities and expectations and set a realistic time frame and price. Recognise a fair price. You only get what you pay for. Asking a contractor to accept unrealistic profit margins can cause problems.

(4) Get the brief right. It's your building. Set a clear brief, to include both an indication of the service you require and the building, and make sure other principal professionals understand your expectations. Don't be afraid to question their suggestions on both organisation and building, but don't attempt to design the building yourself.

(5) Negotiation. Be prepared to negotiate contracts with consultants and contractors in order to keep effective project teams together.

(6) Don't tinker. Changes of mind about use and design once construction has begun cause delays and cost money. Don't change your mind for trivial reasons. If design changes have to be made (and in the real world it does happen) then be systematic about it.

(7) Ask for a thorough site survey. Unexpected ground conditions delay one in two projects.

(8) Keep the project moving. Clients need to communicate their expectations both internally and externally. Make sure there is a clear chain of communication and decision-making within your company so that designers and builders are never held up. Avoid confusion by speaking to them with a single voice.

It is equally important that the industry plays its full part in creating the circumstances necessary for successful projects:

(1) Make clear recommendations. Present clear-cut options and support your design ideas with simple scale models. Do not blind the client with science.

(2) Provide leadership. Roles and responsibilities must be clearly defined and appointments made early to ensure that everyone is on board and moving in the same direction. Complex buildings need effective project management as well as management of the design team and of the site operation.

(3) Provide clear terms of business. The client has a right to know who is responsible for what and how to get redress if something goes wrong. Make sure the team has appropriate indemnity insurance or financial bonding.

(4) Set clear objectives. Only set ambitious objectives, such as completion time, which everyone involved can realistically achieve. As a contractor be prepared to accept financial incentives to back a target date or price.

(5) Safeguard the client's interests. The client doesn't expect to do all the monitoring and checking. Future business depends on client confidence.

(6) Project office. Setting up a project office where the whole project team can work together is very effective and leads to high levels of efficiency and speed.

(7) Innovate slowly and steadily. Build on established answers; original, creative design and construction is difficult, expensive and uncertain. Slow, steady innovation produces consistent benefits.

(8) Speed, quality and efficiency. Speed and punctuality cause good performance in quality, cost and time. Projects with fast planned times both for design and construction finish on time or early. Slow building arises from lack of purpose and momentum, drift, under-resourcing and second-class attention to the project and causes a disappointing outcome all round.

(9) Management control. A comprehensive project strategy and timetable developed at the outset and taken absolutely seriously is essential for success.

(10) Design management. Establish a clear design leader, co-ordinate all design activities and monitor progress, organise and manage the flow of design information. Contractors and specialists have important design roles which should be planned into a total design process from the outset.

(11) Use information technology. New technology — both computer-aided design and information systems — is becoming indispensable to successful projects. CAD helps incorporate specialist elements in the overall design to ensure quality, buildability and speed.

(12) Co-ordinated project information. Use the 'Co-ordinated Project Information' system published by the Committee for Co-ordinated Project Information sponsored by and recommended by ACE, BEC, RIBA and RICS.

(13) Certainty. Do not ask clients to enter into firm lump sum contracts until the design is complete or else the contractor has full authority and responsibility for completing a performance or outline design.

(14) Buildability. Include construction expertise in design decisions and pay particular attention to junctions between elements and systems.

(15) Materials control. Plan the flow of information between suppliers, contractors and designers. Monitor progress in design, manufacture, delivery, distribution on site and assembly. Seek to identify problems early and solve them.

(16) Efficient work on site. Provide clear site management, excellent working conditions and high salaries. Establish two way communications and invest in building team spirit.

(17) Training. Provide training for all workers required to apply new or unusual methods or techniques. This need applies as much to management as to design and construction.

CONCLUSIONS

This chapter provides general guidance on procurement systems. It is vital that each project is considered from basic principles and, just as the design must be tailored to the needs and objectives of the client, so must the procurement system. In order to play a full role in these early crucial decisions, quantity surveyors must understand the strengths and weaknesses of the available systems and know how they match particular projects.

This basic understanding will stand quantity surveyors in good stead as construction becomes more international. The principles learnt in Britain's very flexible construction industry provide a basis for understanding procurement in Europe and indeed world-wide. Languages and cultures vary but the essential links between buildings and people are consistent. The final section of this chapter reviews further literature which describes this basic knowledge.

A GUIDE TO FURTHER STUDY

There is a growing list of writings on procurement systems for building. The Building EDC's guide for clients *Thinking about Building* provides a good starting point for any consideration of how to choose the appropriate system. Skitmore and Marsden's paper takes the EDC ideas further. Nahapiet and Nahapiet provide greater depth in considering client requirements and project characteristics. The Building EDC's two major surveys *Faster building for Industry* and *Faster Building for Commerce* provide well researched practical advice on how to achieve good performance from the available procurement systems. O'Reilly's BRE work *Better Briefing means Better Buildings* is worth looking at as a check list of the issues which should be considered at the outset of projects. Heirs and Pehrson's excellent book provides sound advice on handling the early problem-solving stages of projects. It leads rather nicely into the Engineering Construction EDC's *Guidelines for the Management of Major Projects in the Process Industries* which, despite its title, provides good advice applicable to the procurement and management of all types of construction. This is true also of the Co-ordinating Committee for Project Information's guide to *Co-ordinated Project Information* for Building Works. Its recommendations will help ensure smooth and orderly procurement whichever system is selected. Finally Franks' report for CIOB is a useful general text on procurement.

Advice on specific procurement systems is less common. As far as the traditional system is concerned, most textbooks and advice to practitioners implicitly assume a general contractor approach. Therefore no specific references are given.

There is a large number of writings on the management based systems. However much of this relates to experience in the USA. Two exceptions are Hobson's work and Sidwell's paper which are based on UK projects. They provide useful advice and some hard data on the performance of management contracting. Also Bennett's report for the RICS is based partly on UK experience.

Very little has been published on design and build. Turner's book is most useful while Pain and Bennett's paper is based on a survey of current UK practice and provides a useful statement of the pros and cons of the system.

An important initiative by clients is described in the British Property Federation's Manual. It provides an excellent check-list of responsibilities and duties which must be considered irrespective of which procurement system is used. It also describes an up-to-date version of the general contractor based system.

Finally it is worth reading an academic treatment of the issues involved in procurement systems. Walker's *Project Management in Construction* and Bennett's *Construction Project Management* provide good introductions. They also provide further references into the relevant general management literature.

BIBLIOGRAPHY

Bennett, J. (1983) *Construction Management and the Chartered Quantity Surveyor*. London: Surveyors Publications, RICS.

Bennett, J. (1985) *Construction Project Management*. Sevenoaks: Butterworths.

British Property Federation (1983) *Manual of the BPF System*.

Building EDC (1988) *Faster Building for Commerce*. London: National Economic Development Office.

Building EDC (1983) *Faster Building for Industry*. London: National Economic Development Office.

Building EDC (1985) *Thinking about Building — A Successful Business Customer's Guide to Using the Construction Industry*. London: National Economic Development Office.

CCPI (1987) *Co-ordinated Project Information for Building Works, a Guide with Examples*. Co-ordinating Committee for Project Information.

Engineering Construction EDC (1982) *Guidelines for the Management of Major Projects in the Process Industries*. London: National Economic Development Office.

Franks, J. (1984) *Building Procurement Systems — a Guide to Building Procurement*. The Chartered Institute of Building.

Heirs, B. J. and Pehrson, G. (1977) *The Mind of the Organization*. Plymouth: Harper & Row Ltd.

Hobson, D. (1985) *Management Contracting — A Step in the Right Direction?* London: Surveyors Publication, RICS.

Nahapiet, H. and Nahapiet, J. (1985) A comparison of contractual arrangements for building projects. *Construction Management and Economics*, **3**, 3, pp. 217–31.

O'Reilly, J. J. N. (1987) *Better Briefing means Better Buildings*. Watford: Department of the Environment, Building Research Establishment.

Pain, J. and Bennett, J. (1988) JCT with contractor's design form of contract: a study in use. *Construction Management and Economics*, **6**, 4, pp. 307–37.

Sidwell, A. C. (1983) An evaluation of management contracting. *Construction Management and Economics*, **1**, 1, pp. 47–55.

Skitmore, R. M. and Marsden, D. E. (1988) Which procurement system? Towards a universal procurement selection technique. *Construction Management and Economics*, **6**, 1, pp. 71–89.

Turner, D. (1986) *Design and Build Contract Practice*. Harlow: Longman Group.

Walker, A. (1989) *Project Management in Construction*. Oxford: Blackwell Scientific Publications.

Index